3

Teacher's Resource Book

Yale Mercieca
Lauren White

Contents

BLM INDEX (SEE REPRODUCIBLE RESOURCES)

NELSON TEACHING OBJECTS (NTOS) INDEX

Introduction

Nelson Maths: Australian Curriculum NSW supports the Australian Curriculum content strands *Number and Algebra, Measurement and Geometry* and *Statistics and Probability* F–6, and integrates the proficiency strands of *Understanding, Fluency, Problem Solving* and *Reasoning* throughout the activities and tasks.

This edition of *Nelson Maths* continues the tradition of providing teachers with choice by featuring an array of hands-on tasks and investigative activities. Teachers have the opportunity to tailor a program for their students based on their different learning styles and diverse needs. Each unit provides ideas on how to scaffold students' learning and also provides learning tasks to extend more able students.

Assessment is key in ***Nelson Maths: Australian Curriculum NSW*** – teachers are provided with a variety of opportunities to assess their students' learning for future planning. Teachers can choose assessment methods to best suit individual students, providing them with the opportunity to demonstrate what they have learned. The program provides specific recommendations for future learning experiences, which are linked to the assessment tasks.

Nelson Maths: Australian Curriculum NSW:

- enables teachers to implement focused teaching
- assists students by scaffolding their learning
- improves students' mathematical understandings and thinking
- provides open-ended and stimulating tasks, allowing students to work at their appropriate developmental level
- caters for various and individual learning styles
- develops students' mental computation skills
- provides teachers with choices so they can readily meet the needs of individual students and/or groups of students
- allows for effective grouping of students
- integrates Information and Communication Technology (ICT)
- provides ongoing assessment and opportunities for a variety of assessment types.

Nelson Maths: Australian Curriculum NSW provides between 30 and 33 units of work for each year level. Each unit of work is divided into three Lesson Plans. A Lesson Plan can be taught over one or more teaching sessions, depending on the needs of students. While the Lessons Plans are sequential and should be taught in the order in which they appear in the unit, teachers have the flexibility to complete several activities from a Lesson Plan over a number of days to cater for the learning requirements of the students in their class.

ACRONYMS USED IN *NELSON MATHS: AUSTRALIAN CURRICULUM NSW*

ATC: Assessment Task Card

BLM: Blackline Master

LO: Learning Object (used in Independent Tasks)

ML: Maths Language

NTO: Nelson Teaching Object (able to be used on an interactive whiteboard or personal computer)

PDF: Portable Document Format

Components

TEACHER'S RESOURCE BOOK

The Teacher's Resource Book provides between 30 and 33 units covering the *Australian Curriculum: Mathematics* content strands of *Number and Algebra, Measurement and Geometry* and *Statistics and Probability*.

Each Teacher's Resource Book also provides:

- Assessment Task Cards (with linking Targeted Assessment Task Cards)
- Mid- and End-of-Year Tests (Tests A and B)
- Assessment and Planning BLMs
- Nelson Teaching Objects (NTOs), accessible via a web link
- BLMs, Tests, Answers (Year 3 and up) and Assessment Task Cards as printable PDFs , accessible via the Nelson Primary website.

Each unit is divided into three Lesson Plans. Each Lesson Plan features the following:

At the beginning of each unit the following are listed:
the linking *Australian Curriculum: Mathematics* content strand and NSW *Mathematics K–10 Syllabus* substrand and outcome for that unit.

Maths Language (ML):
In each unit is listed the vocabulary the teacher models and uses in his or her teaching, and the language students are encouraged to use.

Tuning In
This activity orientates students to the focus of the lesson and revises the skills that they will need to use in the lesson. It may provide a link to the previous lesson or previous units. It may also be used as a pre-assessment activity.

Unit 1 **Numbers, Numbers, Numbers**

Number and Algebra
Whole numbers MA2-4NA applies place value to order, read and represent numbers of up to five digits

ML even, hundreds, larger, odd, order, place value, smaller, tens, thousands, units

LESSON PLAN 1

TUNING IN
WHAT'S IN THE JAR?
You will need: beans, four jars or transparent containers
Present students with four jars of beans, with one jar containing 9 beans, the next jar containing 31 beans, the next 53 beans and the final jar 147 beans. Ask, 'Which jar has about 50 beans?' Have students discuss their guess and the reasons for their guess. Invite four students to count the beans in each jar. Record the numbers and tell students that these numbers have something in common that you will ask them about at the end of the lesson.

WHOLE-CLASS INTRODUCTION
EXPLORING ODD AND EVEN NUMBERS
You will need: sets of number cards made from BLM 1 '2-Digit Number Cards', counters
Have students work with a partner. Give each pair some of the number cards made from BLM 1 '2-Digit Number Cards'. Have students count out counters according to their number cards. Explain to students that they are going to determine if they have an odd or even number of counters and that this can be done by grouping counters in twos. If all the counters can be made into groups of two, the number is even. If there is a counter that cannot be grouped with another, the number is odd. Have students determine if their number is odd or even and make a list of the even numbers and the odd numbers.

INDEPENDENT TASKS
Note: Choose from Tasks 1, 2 or 3.
You will need: sticky dots, BLM 1 '2-Digit Number Cards', NTO 3.1 'Ten Frames', Student Book p. 4 'Odd or Even?'

TASK 1: DOT PAIRS
Have students write some 1- and 2-digit numbers, e.g. how old they will be next birthday, the last two digits in their phone number, their street number, the number of students in their class, etc. Students can then make models of the numbers by sticking dots (or drawing dots) in pairs on a sheet of paper and writing the number beside their model. Have students share and discuss their numbers with a partner and then label their models either odd or even.

TASK 2: INTERACTIVE TASK
Give students the number cards from BLM 1 '2-Digit Number Cards' that were not used in the 'Exploring Odd and Even Numbers' activity in the Whole-Class Introduction. Have them use NTO 3.1 'Ten Frames' to determine if the numbers are odd or even.

TASK 3: STUDENT BOOK p. 4 ***'Odd or Even?'***

TEACHING GROUP
You will need: Unifix blocks, BLM 3 'Blank Chart'
FINGERS ON SHOW
- For students who require support, give them some Unifix blocks and have them make the numbers to 10 with two groups of blocks joined, e.g. 4 would have two groups of two blocks. Have students write a list of even numbers and odd numbers. Then have students play a game with a partner whereby they put one hand behind their back and decide how many fingers to hold up. On the count of three, both students

Nelson Maths Australian Curriculum NSW 22 Teacher's Resource Book Year **3**

show their hands. One student will score a point if the total number of fingers shown is even and the other will score a point if it is odd.

HOW DO YOU KNOW IT IS AN EVEN NUMBER?
- For students who require a challenge, give them a copy of BLM 3 'Blank Chart' and have them choose any starting number and fill in the numbers on the chart. Discuss with students the conditions for a number to be even and have them colour all of the even numbers. Have students look at the even numbers and write a rule for how to identify an even number.

REFLECTION
Select from the following to suit your class and their learning outcomes:
- Have students share their work from the Whole-Class Introduction or from the Teaching Groups, and ask, 'How do you know if a number is an even number?' Then invite students to name even numbers that are more than 20, 50, 100, 200, 500 and 1 000. Ask, 'Can you think of a rule for deciding if a number is odd or even?'
- Draw students' attention to the four numbers that you recorded during 'What's in the Jar?' in Tuning In, and ask them what the numbers have in common. Ask, 'How do you know that they are all odd numbers?'

LESSON PLAN 2

TUNING IN
TELL ME ABOUT THE NUMBER
Write a number on the board, e.g. 46, and ask students to think of as many things as they can about the number, e.g. it is an even number, it is made up of 4 tens and 6 ones, it is four less than 50, it is ten more than 36 and so on. Repeat for other numbers.

WHOLE-CLASS INTRODUCTION
WORDS AND NUMERALS
You will need: NTO 3.2 'Extended Notation: Number Cards', small whiteboards
Present NTO 3.2 'Extended Notation: Number Cards', selecting the random number to be shown. Have students read the number then invite a student to show the number with the number cards. Explain that we can see the number written as a numeral and shown with number cards. Ask, 'How can you write the number in words?' Have students say the number and then ask them to write the number in words on their whiteboards. To support students, write the words 'thousand', 'hundred', 'ninety', 'eighty', 'seventy', 'sixty', 'fifty', 'forty', 'thirty' and 'twenty' on the board or on cards. Repeat for other numbers.

INDEPENDENT TASKS
Note: Choose from Tasks 1, 2 or 3.
You will need: BLM 4 'Blank Cards', NTO 3.3 'Six-Sided Dice', Student Book p. 5 'Number Sort'

TASK 1: CARD GAME
Give each student two copies of BLM 4 'Blank Cards'. Have them make pairs of cards with numerals on half of the cards and the matching numerals in words on the other half. Students combine with a partner to play 'Memory', whereby all of the cards are placed face-down and students take it in turns to turn over two cards to find a pair. Alternatively, students could play 'Go Fish' or 'Snap'.

TASK 2: INTERACTIVE TASK
Have students work with a partner using NTO 3.3 'Six-Sided Dice'. Before rolling the dice, students decide if they will form an odd or even number. They roll four dice and rearrange them to form a 4-digit odd or even number, which they record using words. Each time they meet the requirement, they score a point.

TASK 3: STUDENT BOOK p. 5 ***'Number Sort'***

TEACHING GROUP
You will need: NTO 3.3 'Six-Sided Dice', five dice for each student in the group
STEP BY STEP
- For students who require support, begin with 2-digit numbers and build up to 4-digit numbers. Present NTO 3.3 'Six-Sided Dice' and select two dice to randomly generate two digits. Have students write the two possible 2-digit numbers and determine if the numbers are odd or even. Have students write the numbers in words. Have students add 100 to the numeric form of their numbers and ask them to read the numbers and write them in words. Then have students add 1 000 to the numeric form of their numbers and have them read the numbers and write them in words. Repeat the process by rolling the two dice again.

Unit **1** Numbers, Numbers, Numbers 23

Whole-Class Introduction
The teacher introduces the main focus of the lesson to the whole class. Through carefully planned activities and questioning, students are introduced to new skills and understandings, which they will develop further throughout the other components of the lesson. Students who need further support or extension in a Teaching Group can be identified at this time.

Independent Tasks
Students work in small groups, pairs or individually to further strengthen their understandings and skills. The teacher may choose to have students complete an open-ended task, an interactive Learning Object, NTO or ICT task and/or the accompanying Student Book page. Depending on the learning needs of students, one or more of these tasks may be chosen.

Reflection

The class regroups as a whole to reflect upon and celebrate their learning. This part of the lesson promotes the use of mathematical language, enables the teacher to gain valuable insights into students' mathematical thinking, and reinforces the main ideas and concepts taught during the lesson.

Home Tasks

These tasks are designed to be undertaken at home. They allow students to share what they have learned in class and to reinforce their mathematical understandings with parents and/or carers. Home Tasks are provided for students in Year 3 and above.

FOUR CATEGORIES

- For students who require a challenge, have them work with higher numbers. Give each student five dice that they use to form four 5-digit numbers: the highest odd number, the highest even number, the lowest odd number and the lowest even number. Have students write their numbers in words and compare with the group. A point is given to the student who is able to form the highest or lowest in each category, and a point deducted if they are unable to form a number for any of the categories. Have students continue to roll the dice and form more numbers.

REFLECTION

Select from the following to suit your class and their learning outcomes:

- Have students discuss any difficulties they encountered when writing numbers in words and ask, 'How did you know that you had written the number correctly?'
- Create a chart of students' strategies for writing numbers in words. Display the chart in the classroom.

LESSON PLAN 3

TUNING IN

POPCORN!

Explain to students that they are going to play a game in which you will call out numbers, and if the number is odd they crouch down, but if the number is even, they jump up and say: 'Popcorn'. Have students stand, and begin by calling out 1- and 2-digit numbers. As the game progresses, increase the numbers to 3- or 4-digit numbers.

WHOLE-CLASS INTRODUCTION

MOVING NUMBERS

You will need: number cards made from BLM 2 'Mixed Number Cards'

Have students sit on chairs in a circle and give each student a number card made from BLM 2 'Mixed Number Cards'. Explain to students that, if their number meets the criteria mentioned, they need to stand up and swap chairs with someone else. Begin by giving criteria such as 'more than 500', 'odd number', '7 in the tens place', '4 units' or '4-digit number'. When students have had practice, say 'even number' and take away a chair as students are moving around so that one student is left in the centre of the circle. That student must give some criteria and try to find an empty chair to sit on while the others are moving.

INDEPENDENT TASKS

Note: Choose from Tasks 1, 2 or 3.

You will need: NTO 3.4 'Playing Cards', Student Book p. 6 'True Statement'

TASK 1: BEAT THE TEACHER

Model the following game, then have groups of three students play the game using NTO 3.4 'Playing Cards' on individual computers. One student must act as the teacher, while the other two students make the 4-digit number. In pairs, have students draw four boxes joined together in a horizontal line. Explain to students that the following activity involves them making the largest 4-digit number they can and both students need to agree where to place the numbers. Using NTO 3.4 'Playing Cards', generate a card and students decide in which box to place the number. (You must also choose a 4-digit number, but do not reveal it to the class.) The game continues until four cards have been generated and students have placed the numbers into boxes. Say your 4-digit number to the class. If a pair's number is lower than your number, they score 1 point, and if it is equal to your number, they score 3 points, but if it is higher than your number, they score 5 points. You score 10 points if you beat all the students. Note: when students are familiar with the game, make it more challenging by deciding that the number must be odd or even.

TASK 2: INTERACTIVE TASK

Have students work with a partner. Using NTO 3.4 'Playing Cards', the first student draws four cards that they rearrange to make the largest odd or even number they can. Their partner then draws four cards and tries to make a larger odd or even number.

TASK 3: STUDENT BOOK p. 6 *'True Statement'*

TEACHING GROUP

You will need: dice

BONGEL

- For students who require support, have them work with smaller numbers. Have students work with two or three dice, playing with a partner or in a small group. Each student rolls the dice to form the largest

Nelson Maths Australian Curriculum NSW 24 Teacher's Resource Book Year 3

number they can. The student with the largest number in this round writes the letter 'B'. Students roll the dice again to form the largest number they can, and each time a student wins a round, they write a letter. The first student to form the word 'BONGEL' wins.

HOW MANY NUMBERS?

- For students who require a challenge, have them work with larger numbers. Give students five dice to roll and have them form and record the smallest number they can. Students roll the dice again to form a 5-digit number that is more than the previous number. Students continue to roll the dice and each time they must form and record a number that is larger than the one before; if they cannot, they must stop. Have students compare how many numbers they formed.

REFLECTION

Select from the following to suit your class and their learning outcomes:

- Have students share the strategies they used to help them win the games they played.
- Draw four boxes on the board and tell students that you want them to form the largest number they can. Roll a 10-sided dice, and ask, 'Which box would you put the number in? Why do you think that is a good choice?'

Home Task

- Have students look at the newspaper (hard copy or online) for examples of numbers used in the media. Have them find examples of 4-digit numbers and what they are used for in daily life.

Assessment

- Have students complete **Student Assessment p. 7**.
- Review with students **Assessment Task Card 3.1**.

During the three lessons:

- Observe which students are able to read and write numbers in word and numeric form during 'Card Game' in Lesson Plan 2, Independent Tasks, Task 1, and mark on a class list.
- Make note of students completing the scaffolding tasks or the more challenging activities of the Teaching Groups.
- Review Student Book pages and make notes of areas of difficulty.

Recommendations for Future Learning

Specific to Student Assessment p. 7; if the student is experiencing difficulty:

Q 1 Have the student use counters to explore which numbers can make groups of two without any leftover counters. Have the student mark these even numbers on a 100 chart and identify patterns associated with even numbers.

Q 2–3 Include numbers in word form in weekly spelling lists.

Q 4 Have the student revisit what makes a number even or odd and have them model 2-, 3- and 4-digit numbers using MAB. Have them order numbers by comparing models.

If the student has not achieved the recommended skills for this unit:

1. See **Assessment Task Card 3.1** for specific recommendations.
2. Have the student work with 2- and 3-digit numbers in any of the listed activities or tasks prior to moving to 4-digit numbers.
3. Review *Nelson Maths: Australian Curriculum NSW Year 2* Unit 1.

If the student has achieved the recommended skills and these skills are firmly established, consider:

1. Having the student complete *Nelson Maths Building Mental Strategies Skill Book Year 4*, pp. 10–11, to reinforce mental strategies with 4- and 5-digit numbers.
2. Moving forward to *Nelson Maths: Australian Curriculum NSW Year 4* Unit 1.
3. Extending the student in any of the listed activities or tasks by using larger numbers.

Unit **1** Numbers, Numbers, Numbers 25

Teaching Group

The teacher works with a small number of 'like-needs' students to teach level-appropriate concepts. The aim of this session is for the teacher to assist students at their point of need and then move them on. This is also a time to make anecdotal notes about individual students. After 10 to 20 minutes, students work on related independent tasks, while the teacher 'roves' amongst the class. The first task provided is for those students requiring further assistance, and the second task is for students who need more challenging experiences.

Assessment

Specific assessment advice is provided for teachers at the conclusion of each unit. Students should complete the Student Assessment page in the Student Book. The purpose of this page is to monitor students' understandings on the specific topic taught in the unit.

An Assessment Task Card (ATC) for each unit is also provided. This card most often provides a hands-on, investigative type of assessment task, which is less 'test focused' than the Student Assessment page. It can be used individually, with small groups or in a whole-class situation.

Recommendations for Future Learning

Each unit lists specific recommendations for future learning for both the Student Assessment page in the Student Book and the Assessment Task Card. Note: Test A provides a six-month test-based assessment covering all of the content strands taught in the first half-year period. Test B assesses content covering the whole year. Tests A and B provide teachers with a comprehensive assessment of their students' achievement levels and mathematical understandings.

STUDENT BOOK

The Student Book features engaging tasks that students can complete independently.

The Student Book includes:

Three pages of activities for each unit (one Student Book page per Lesson Plan)

The linking NSW *Mathematics K–10 Syllabus* substrand and outcome for that unit

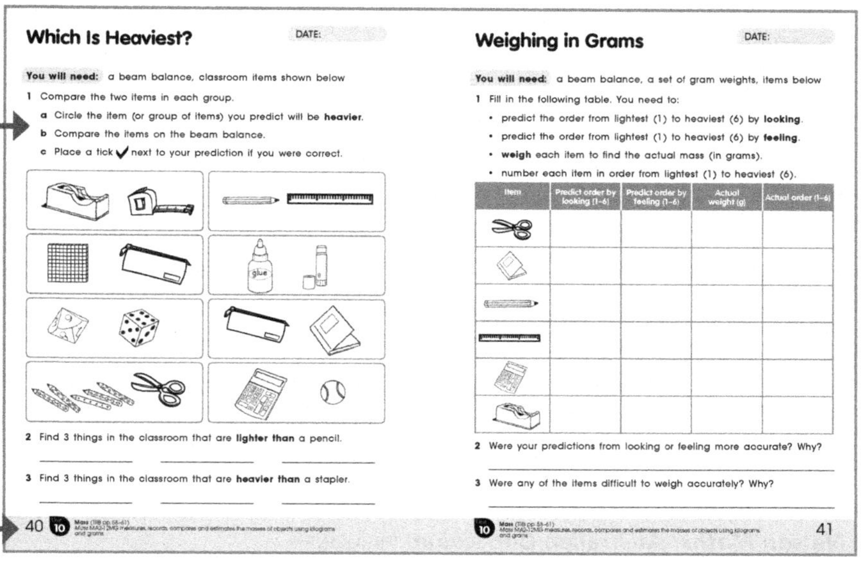

Which Is Heaviest? DATE:

You will need: a beam balance, classroom items shown below

1 Compare the two items in each group.

a Circle the item (or group of items) you predict will be **heavier**.

b Compare the items on the beam balance.

c Place a tick ✓ next to your prediction if you were correct.

2 Find 3 things in the classroom that are **lighter than** a pencil.

3 Find 3 things in the classroom that are **heavier than** a stapler.

40 10

Weighing in Grams DATE:

You will need: a beam balance, a set of gram weights, items below

1 Fill in the following table. You need to:

- predict the order from lightest (1) to heaviest (6) by **looking**.
- predict the order from lightest (1) to heaviest (6) by **feeling**.
- **weigh** each item to find the actual mass (in grams).
- number each item in order from lightest (1) to heaviest (6).

Item	Predict order by looking (1–6)	Predict order by feeling (1–6)	Actual weight (g)	Actual order (1–6)

2 Were your predictions from looking or feeling more accurate? Why?

3 Were any of the items difficult to weigh accurately? Why?

10 41

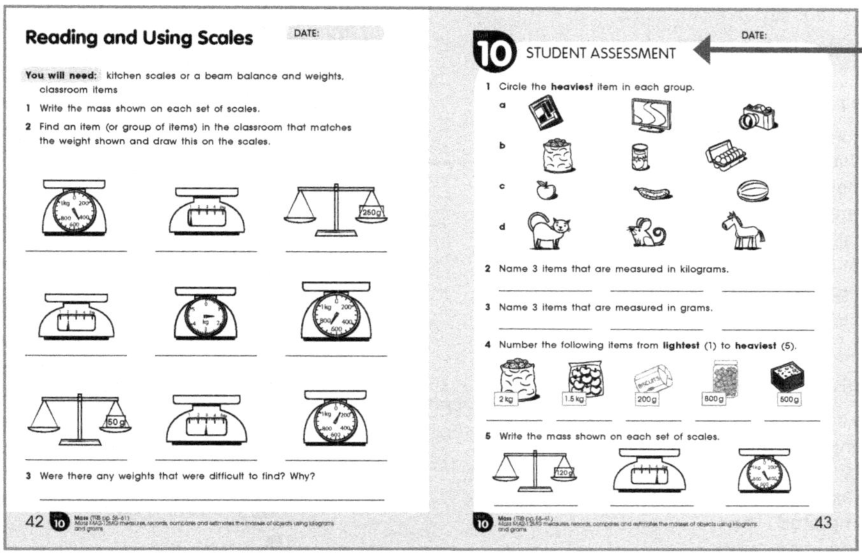

Reading and Using Scales DATE:

You will need: kitchen scales or a beam balance and weights, classroom items

1 Write the mass shown on each set of scales.

2 Find an item (or group of items) in the classroom that matches the weight shown and draw this on the scales.

3 Were there any weights that were difficult to find? Why?

42 10

10 STUDENT ASSESSMENT DATE:

1 Circle the **heaviest** item in each group.

2 Name 3 items that are measured in kilograms.

3 Name 3 items that are measured in grams.

4 Number the following items from **lightest** (1) to **heaviest** (5).

2 kg 1.5 kg 200 g 800 g 500 g

5 Write the mass shown on each set of scales.

10 43

One Student Assessment page per unit. This page enables both student and teacher to monitor and assess an individual student's mathematical understandings throughout the units of work. Students can complete the page independently or with teacher assistance. The resulting data will assist teachers with future planning and can be used as part of a student's portfolio (for teacher–parent interviews).

Also included at the back of the Student Book is a Maths Glossary.

NELSON TEACHING OBJECTS AND LEARNING OBJECTS

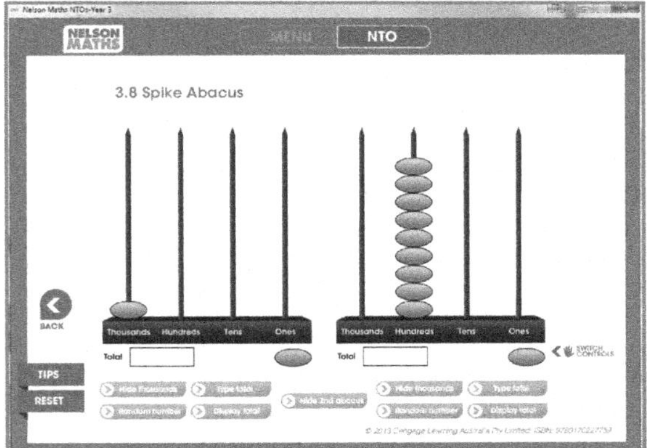

Nelson Maths: Australian Curriculum integrates Information and Communication Technology (ICT) into the program. It provides up to 59 Nelson Teaching Objects (NTOs) for use on interactive whiteboards, individual computers and interactive LCD screens. The NTOs:

- illustrate mathematical concepts explicitly
- engage students actively in their own learning
- scaffold student learning
- require students to use their mathematical understandings in an engaging and meaningful context.

Also referred to are the pedagogically sound interactive Learning Objects (LOs) created by Education Services Australia (ESA). These are incorporated into Independent Tasks in the unit Lesson Plans. Teachers can access the LOs at:

http://www.scootle.edu.au/ec/p/home

Assessment and Planning

Assessment is a vital and ongoing part of the teaching and learning experience. Teachers need to continually monitor and assess students so their learning needs can be met and targeted when planning future educational experiences.

Teachers need to assess students' mathematical understandings (verbal and recorded), fluency, problem-solving and reasoning abilities, confidence and ability to work with others.

Assessment is ongoing, but it can be specifically targeted:

- at the middle or end of the school year (Test A and Test B)
- at the beginning of a unit (Tuning In, previous Assessment Task Card)
- at the end of a unit (Student Assessment page, Assessment Task Card).

Assessment can occur daily during mathematics sessions as part of the Whole-Class Introduction, Teaching Groups and Independent Tasks, while roving, and during Reflection.

It is not only the assessment that is important, but also what occurs after the assessment. ***Nelson Maths: Australian Curriculum*** provides comprehensive and specific suggestions for students requiring further assistance as well as for those who require more challenging experiences.

ASSESSMENT RESOURCES

Nelson Maths: Australian Curriculum includes the following assessment resources:

Student Assessment page (one page per unit)

This page links to Recommendations for Future Learning in the Teacher's Resource Book to guide teachers with strategic or targeted responses to the specific learning needs of individual students. The Student Assessment page is finely targeted – the questions on this page assess particular skills, enabling teachers to explicitly identify where a student may be struggling. Teachers can then respond to that specific need accordingly.

DATE:

16 STUDENT ASSESSMENT

1 a How many seconds are there in 1 minute? ________
b How many minutes are there in 1 hour? ________

2 Write the following times in seconds into the correct box.

34 seconds 55 seconds 75 seconds
19 seconds 63 seconds 2 seconds
81 seconds 40 seconds 100 seconds

Less than 1 minute	About a minute	More than 1 minute

3 Write the times shown on these clocks.

4 Draw the missing hands on each clock:

10:15 10:35 9:20 9:50 7:02

5 Think of two activities we measure in:
a seconds: ________ ________
b minutes: ________ ________
c hours: ________ ________

16 67

Assessment Task Cards (one for each of the 30 to 33 units)

These cards have been designed to be used with individuals, in small groups or as a whole-class assessment task. Teachers may choose to withdraw individual students to assess their understandings, work with a small group or instruct the whole class on the assessment task to be completed. The Assessment Task Cards are an alternative to pen-and-paper tests and provide some students with a more accurate way to demonstrate what they understand.

The Assessment Task Cards can be used before commencing a unit to assess students' prior knowledge, but are mostly intended to be used after the completion of a unit to assess students' understandings.

Each Assessment Task Card has a linking Targeted Assessment Task Card, which provides suggestions for where to take students next as well as specific recommendations for future learning.

All Assessment Task Cards link to the relevant NSW *Mathematics K–10 Syllabus* substrand and outcome. The cards are designed to be printed, laminated and stored for repeated use.

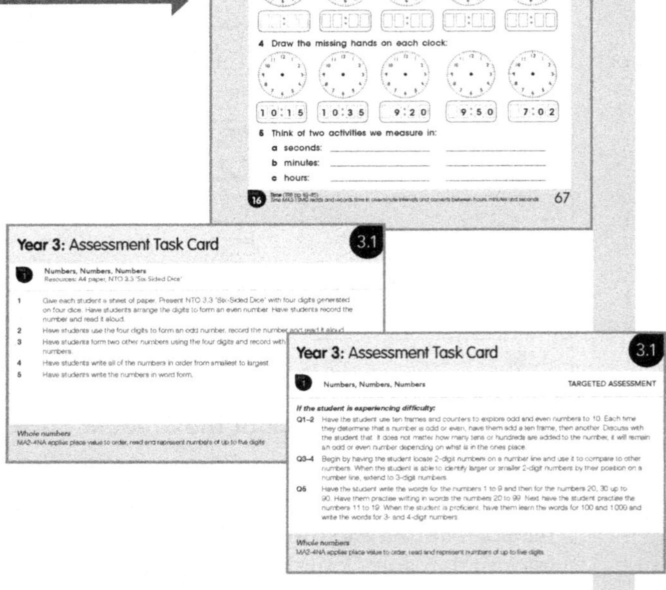
Year 3: Assessment Task Card 3.1

1 Numbers, Numbers, Numbers
Resources: A4 paper, NTO 3.3 'Six-Sided Dice'

1 Give each student a sheet of paper. Present NTO 3.3 'Six-Sided Dice' with four digits generated on four dice. Have students arrange the digits to form an even number. Have students record the number and read it aloud.
2 Have students use the four digits to form an odd number, record the number and read it aloud.
3 Have students form two other numbers using the four digits and record with… numbers.
4 Have students write all of the numbers in order from smallest to largest.
5 Have students write the numbers in word form.

Whole numbers
MA2-4NA applies place value to order, read and represent numbers of up to five digits

Year 3: Assessment Task Card 3.1

1 Numbers, Numbers, Numbers TARGETED ASSESSMENT

If the student is experiencing difficulty:

Q1–2 Have the student use ten frames and counters to explore odd and even numbers to 10. Each time they determine that a number is odd or even, have them add a ten frame, then another. Discuss with the student that it does not matter how many tens or hundreds are added to the number, it will remain an odd or even number depending on what is in the ones place.

Q3–4 Begin by having the student locate 2-digit numbers on a number line and use it to compare to other numbers. When the student is able to identify larger or smaller 2-digit numbers by their position on a number line, extend to 3-digit numbers.

Q5 Have the student write the words for the numbers 1 to 9 and then for the numbers 20, 30 up to 90. Have them practise writing in words the numbers 20 to 99. Next have the student practise the numbers 11 to 19. When the student is proficient, have them learn the words for 100 and 1000 and write the words for 3- and 4-digit numbers.

Whole numbers
MA2-4NA applies place value to order, read and represent numbers of up to five digits

Mid- and End-of-Year Tests (Tests A and B)

These two paper-based tests assess students' understandings across the content strands. They are designed for mid- and end-of-year assessment. However, teachers may choose to use, for example, Year 3 Test B at the beginning of Year 4.

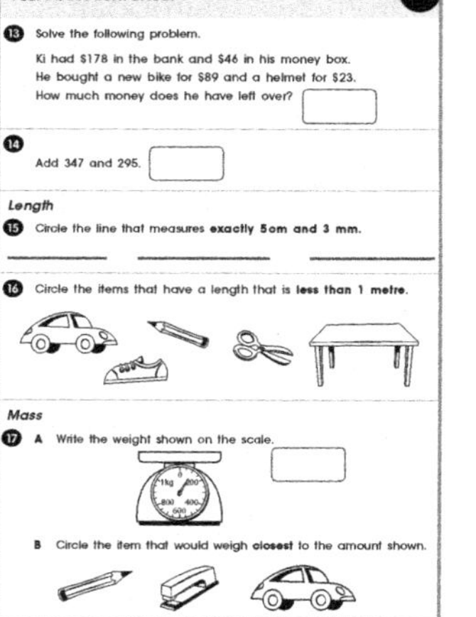
Test A: Student Sheet 3.A

13 Solve the following problem.
Ki had $178 in the bank and $46 in his money box.
He bought a new bike for $89 and a helmet for $23.
How much money does he have left over?

14 Add 347 and 295.

Length
15 Circle the line that measures **exactly 5 cm and 3 mm.**

16 Circle the items that have a length that is **less than 1 metre.**

Mass
17 A Write the weight shown on the scale.
B Circle the item that would weigh **closest** to the amount shown.

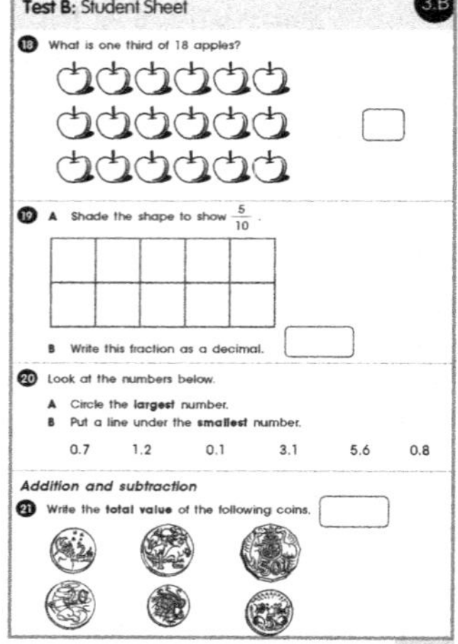
Test B: Student Sheet 3.B

18 What is one third of 18 apples?

19 A Shade the shape to show $\frac{5}{10}$.
B Write this fraction as a decimal.

20 Look at the numbers below.
A Circle the **largest** number.
B Put a line under the **smallest** number.
0.7 1.2 0.1 3.1 5.6 0.8

Addition and subtraction
21 Write the **total value** of the following coins.

The Classroom Environment

What teachers know about learning:

- Students learn in different ways.
- Students learn from and with others.
- Students should have an opportunity to record in different ways.
- Students can follow different pathways to reach the same learning milestones.
- Students need to be encouraged to take risks in their learning.
- Students learn by being actively involved.
- Students learn by teaching others.
- Students need time to grasp the concepts introduced; that is, time for thinking, reflecting, reacting and sharing the processes and answers in mathematics.
- Students learn when they feel good about themselves – healthy self-esteem is critical.

It is important that the classroom environment:

- is stimulating and engaging and enables students to become fully immersed in learning.
- incorporates a variety of rich resources that engage students in making mathematical connections.
- is rich in the language of mathematics.
- provides students with access to calculators.
- provides students with access to Information Technology, which is an integral part of our world.
- has rich displays where students can make metacognitive links to their mathematics learning.
- is positive and encouraging and enables students to take risks in their learning.

These ideas will help to create a stimulating mathematics classroom:

- Provide step-by-step charts so students can work independently to resolve problems.
- Display students' work. Allow time to reflect upon and view completed mathematical tasks.
- Ensure students are aware of the mathematical material available to them.
- Display individual students' expertise so others can learn from them.
- Encourage students to work cooperatively in small groups.
- Integrate mathematics across the strands.

- Use charts to display helpful information and promote mathematical language.
- Include students in decisions about upcoming mathematical learning.
- Encourage peer tutoring.
- Integrate ICT into the everyday learning environment.
- Illustrate mathematics in real-world contexts.

Year 3 Overview

Unit	AC Strand/NSW Mathematics K–10 Syllabus Substrand & Outcome/AC Content Description & Code	NTOs	BLMs
Unit 1 Numbers, Numbers, Numbers	**Number and Algebra** *Whole numbers* MA2-4NA applies place value to order, read and represent numbers of up to five digits Investigate the conditions required for a number to be odd or even and identify odd and even numbers (ACMNA051) AC Recognise, model, represent and order numbers to at least 10 000 (ACMNA052) AC	3.1 Ten Frames 3.2 Extended Notation: Number Cards 3.3 Six-Sided Dice 3.4 Playing Cards	BLM 1 2-Digit Number Cards BLM 2 Mixed Number Cards BLM 3 Blank Chart BLM 4 Blank Cards
Unit 2 Numbers to 10 000	**Number and Algebra** *Whole numbers* MA2-4NA applies place value to order, read and represent numbers of up to five digits Recognise, model, represent and order numbers to at least 10 000 (ACMNA052) AC	3.5 Place-Value Dice 3.6 Number Line	BLM 2 Mixed Number Cards BLM 5 Digit Cards
Unit 3 More About Numbers to 10 000	**Number and Algebra** *Whole numbers* MA2-4NA applies place value to order, read and represent numbers of up to five digits Recognise, model, represent and order numbers to at least 10 000 (ACMNA052) AC	3.7 Modelling with MAB 3.8 Spike Abacus	BLM 6 4-Digit Number Cards BLM 7 MAB BLM 8 Place-Value Mat BLM 9 Zero's the Go!
Unit 4 Length	**Measurement and Geometry** *Length* MA2-9MG measures, records, compares and estimates lengths, distances and perimeters in metres, centimetres and millimetres, and measures, compares and records temperatures Measure, order and compare objects using familiar metric units of length, mass and capacity (ACMMG061) AC		BLM 10 Grid Paper BLM 11 Curved lines

Unit	AC Strand/NSW Mathematics K–10 Syllabus Substrand & Outcome/AC Content Description & Code	NTOs	BLMs
Unit 5 Mental Strategies for Addition	**Number and Algebra** *Addition and subtraction* MA2-5NA uses mental and written strategies for addition and subtraction involving two-, three-, four- and five-digit numbers Recall addition facts for single-digit numbers and related subtraction facts to develop increasingly efficient mental strategies for computation (ACMNA055) AC	3.9 100 Chart 3.10 Card Flip	BLM 5 Digit Cards BLM 12 Sports Equipment Prices BLM 13 Addition Cards BLM 14 100 Chart BLM 15 Number Cards 1–20 BLM 16 Near Doubles BLM 17 Blank Dominoes
Unit 6 Addition	**Number and Algebra** *Addition and subtraction* MA2-5NA uses mental and written strategies for addition and subtraction involving two-, three-, four- and five-digit numbers Recognise and explain the connection between addition and subtraction (ACMNA054) AC	3.7 Modelling with MAB 3.10 Card Flip 3.11 Calculator	BLM 1 2-Digit Number Cards BLM 5 Digit Cards BLM 8 Place-Value Mat BLM 13 Addition Cards BLM 15 Number Cards 1–20 BLM 18 Check This BLM 19 Addition Boxes
Unit 7 Position	**Measurement and Geometry** *Position* MA2-17MG uses simple maps and grids to represent position and follow routes, including using compass directions Create and interpret simple grid maps to show position and pathways (ACMMG065) AC		BLM 10 Grid Paper BLM 20 Directions BLM 21 Shopping Centre Map
Unit 8 Place Value	**Number and Algebra** *Whole numbers* MA2-4NA applies place value to order, read and represent numbers of up to five digits Apply place value to partition, rearrange and regroup numbers to at least 10 000 to assist calculations and solve problems (ACMNA053) AC	3.2 Extended Notation: Number Cards 3.7 Modelling with MAB 3.12 Stopwatch	BLM 6 4-Digit Number Cards BLM 7 MAB BLM 8 Place-Value Mat BLM 22 Banknotes BLM 23 4-Digit Number Expanders BLM 24 Phone Number Hunt

Unit	AC Strand/NSW Mathematics K–10 Syllabus Substrand & Outcome/AC Content Description & Code	NTOs	BLMs
Unit 9 More About Place Value	**Number and Algebra** *Whole numbers* MA2-4NA applies place value to order, read and represent numbers of up to five digits Apply place value to partition, rearrange and regroup numbers to at least 10 000 to assist calculations and solve problems (ACMNA053) AC		BLM 6 4-Digit Number Cards BLM 8 Place-Value Mat BLM 25 Number Hieroglyphs BLM 26 Place-Value Ladders BLM 27 Dominoes
Unit 10 Mass	**Measurement and Geometry** *Mass* MA2-12MG measures, records, compares and estimates the masses of objects using kilograms and grams Measure, order and compare objects using familiar metric units of length, mass and capacity (ACMMG061) AC		BLM 28 Mass Cards
Unit 11 Our Community – Data	**Statistics and Probability** *Data* MA2-18SP selects appropriate methods to collect data, and constructs, compares, interprets and evaluates data displays, including tables, picture graphs and column graphs Identify questions or issues for categorical variables. Identify data sources and plan methods of data collecting and recording (ACMSP068) AC Collect data, organise into categories and create displays using lists, tables, picture graphs and simple column graphs, with and without the use of technologies (ACMSP069) AC Interpret and compare data displays (ACMSP070) AC	3.13 Data Collection Table 3.14 Column Graph 3.15 Picture Graph 3.16 Audit Process	BLM 10 Grid Paper BLM 29 Data Collection Table BLM 30 Graph It!
Unit 12 Mental Strategies for Subtraction	**Number and Algebra** *Addition and subtraction* MA2-5NA uses mental and written strategies for addition and subtraction involving two-, three-, four- and five-digit numbers Recall addition facts for single-digit numbers and related subtraction facts to develop increasingly efficient mental strategies for computation (ACMNA055) AC	3.6 Number Line 3.7 Modelling with MAB 3.9 100 Chart 3.10 Card Flip 3.11 Calculator	BLM 1 2-Digit Number Cards BLM 14 100 Chart BLM 15 Number Cards 1–20 BLM 31 Counting by 10 Cards BLM 32 Number Lines

Unit	AC Strand/NSW Mathematics K–10 Syllabus Substrand & Outcome/AC Content Description & Code	NTOs	BLMs
Unit 13 Subtraction	**Number and Algebra** *Addition and subtraction* MA2-5NA uses mental and written strategies for addition and subtraction involving two-, three-, four- and five-digit numbers Recognise and explain the connection between addition and subtraction (ACMNA054) AC	3.7 Modelling with MAB 3.10 Card Flip 3.11 Calculator	BLM 1 2-Digit Number Cards BLM 8 Place-Value Mat BLM 33 Subtraction Cards BLM 34 Bingo Boards BLM 31 Counting by 10 Cards
Unit 14 Connections Between Addition and Subtraction	**Number and Algebra** *Addition and subtraction* MA2-5NA uses mental and written strategies for addition and subtraction involving two-, three-, four- and five-digit numbers Recognise and explain the connection between addition and subtraction (ACMNA054) AC Recall addition facts for single-digit numbers and related subtraction facts to develop increasingly efficient mental strategies for computation (ACMNA055) AC	3.1 Ten Frames 3.7 Modelling with MAB 3.11 Calculator 3.17 Ten-Sided Dice	BLM 1 2-Digit Number Cards BLM 3 Blank Chart BLM 5 Digit Cards BLM 15 Number Cards 1–20 BLM 33 Subtraction Cards BLM 35 Fact Family Houses BLM 36 Missing Number Maze
Unit 15 Solving Addition and Subtraction Problems	**Number and Algebra** *Addition and subtraction* MA2-5NA uses mental and written strategies for addition and subtraction involving two-, three-, four- and five-digit numbers Recognise and explain the connection between addition and subtraction (ACMNA054) AC	3.7 Modelling with MAB 3.11 Calculator	BLM 4 Blank Cards BLM 5 Digit Cards BLM 8 Place-Value Mat BLM 12 Sports Equipment Prices BLM 37 3-Digit Number Cards BLM 38 Letter Codes BLM 39 Jungle Map BLM 40 Worded Problems BLM 41 Robot Parts

Unit	AC Strand/NSW Mathematics K–10 Syllabus Substrand & Outcome/AC Content Description & Code	NTOs	BLMs
Unit 16 Time	**Measurement and Geometry** *Time* MA2-13MG reads and records time in one-minute intervals and converts between hours, minutes and seconds Tell the time to the minute and investigate the relationship between units of time (ACMMG062) AC	3.12 Stopwatch 3.18 Clocks	BLM 42 Typing to the Minute BLM 43 Time Cards BLM 44 Match the Time BLM 45 Blank Clock Faces BLM 46 Making a Clock BLM 47 *PowerPoint* Clocks
Unit 17 Angles	**Measurement and Geometry** *Angles* MA2-16MG identifies, describes, compares and classifies angles Identify angles as measures of turn and compare angle sizes in everyday situations (ACMMG064) AC	3.19 Rotating Lines	BLM 3 Blank Chart BLM 48 Label the Lines
Unit 18 Mental Strategies for Multiplication	**Number and Algebra** *Multiplication and division* MA2-6NA uses mental and informal written strategies for multiplication and division Recall multiplication facts of two, three, five and ten and related division facts (ACMNA056) AC	3.7 Modelling with MAB 3.9 100 Chart 3.10 Card Flip 3.17 Ten-Sided Dice 3.20 Counters	BLM 4 Blank Cards BLM 5 Digit Cards BLM 31 Counting by 10 Cards BLM 49 Rocket Race
Unit 19 More About Mental Strategies for Multiplication	**Number and Algebra** *Multiplication and division* MA2-6NA uses mental and informal written strategies for multiplication and division Recall multiplication facts of two, three, five and ten and related division facts (ACMNA056) AC	3.7 Modelling with MAB 3.11 Calculator 3.17 Ten-Sided Dice	BLM 5 Digit Cards BLM 50 4 in a Row BLM 51 Multiplication Fact Cards

Unit	AC Strand/NSW Mathematics K–10 Syllabus Substrand & Outcome/AC Content Description & Code	NTOs	BLMs
Unit 20 Chance	**Statistics and Probability** *Chance* MA2-19SP describes and compares chance events in social and experimental contexts Conduct chance experiments, identify and describe possible outcomes and recognise variation in results (ACMSP067) AC	3.3 Six-Sided Dice 3.21 Coin Flip 3.22 Spinner	BLM 4 Blank Cards BLM 52 Car Race BLM 53 Spinners
Unit 21 Patterns	**Number and Algebra** *Patterns and algebra* MA2-8NA generalises properties of odd and even numbers, generates number patterns, and completes simple number sentences by calculating missing values Describe, continue and create number patterns resulting from performing addition and subtraction (ACMNA060) AC	3.6 Number Line 3.9 100 Chart	BLM 3 Blank Chart BLM 14 100 Chart BLM 55 Start at… Cards BLM 56 Number Pattern Rules BLM 57 Chinese Zodiac
Unit 22 Multiplication	**Number and Algebra** *Multiplication and division* MA2-6NA uses mental and informal written strategies for multiplication and division Represent and solve problems involving multiplication using efficient mental and written strategies and appropriate digital technologies (ACMNA057) AC	3.7 Modelling with MAB 3.11 Calculator	BLM 1 2-Digit Number Cards BLM 4 Blank Cards BLM 5 Digit Cards BLM 10 Grid Paper BLM 12 Sports Equipment Prices BLM 58 Multiplication Cards BLM 59 Multiplication Pictures BLM 60 Think Board
Unit 23 Area	**Measurement and Geometry** *Area* MA2-10MG measures, records, compares and estimates areas using square centimetres and square metres Measure, order and compare objects using familiar metric units of length, mass and capacity (ACMMG061) AC	3.23 Area of Shapes	BLM 10 Grid Paper BLM 61 Area of Rectangles BLM 62 Square Dot Paper BLM 63 Tatami Mats

Unit	AC Strand/NSW Mathematics K–10 Syllabus Substrand & Outcome/AC Content Description & Code	NTOs	BLMs
Unit 24 Division	**Number and Algebra** *Multiplication and division* MA2-6NA uses mental and informal written strategies for multiplication and division Recall multiplication facts of two, three, five and ten and related division facts (ACMNA056) AC	3.6 Number Line 3.7 Modelling with MAB 3.10 Card Flip	BLM 1 2-Digit Number Cards BLM 5 Digit Cards BLM 59 Multiplication Pictures BLM 60 Think Board BLM 64 Division Problems
Unit 25 More About Division	**Number and Algebra** *Multiplication and division* MA2-6NA uses mental and informal written strategies for multiplication and division Recall multiplication facts of two, three, five and ten and related division facts (ACMNA056) AC	3.6 Number Line 3.11 Calculator 3.17 Ten-Sided Dice 3.20 Counters	BLM 5 Digit Cards BLM 10 Grid Paper BLM 17 Blank Dominoes BLM 51 Multiplication Fact Cards BLM 58 Multiplication Cards BLM 64 Division Problems BLM 65 Array Cards BLM 66 Multiply or Divide
Unit 26 Fractions	**Number and Algebra** *Fractions and decimals* MA2-7NA represents, models and compares commonly used fractions and decimals Model and represent unit fractions including $\frac{1}{2}, \frac{1}{4}, \frac{1}{3}, \frac{1}{5}$ and their multiples to a complete whole (ACMNA058) AC	3.24 Fraction Circles	BLM 10 Grid Paper BLM 67 Fraction Wall BLM 68 Fraction Cards
Unit 27 More About Fractions	**Number and Algebra** *Fractions and decimals* MA2-7NA represents, models and compares commonly used fractions and decimals Model and represent unit fractions including $\frac{1}{2}, \frac{1}{4}, \frac{1}{3}, \frac{1}{5}$ and their multiples to a complete whole (ACMNA058) AC	3.6 Number Line	BLM 67 Fraction Wall BLM 70 Digital Fraction Wall BLM 71 Horse Race BLM 72 Cars in Our Car Park

Unit	AC Strand/NSW Mathematics K–10 Syllabus Substrand & Outcome/AC Content Description & Code	NTOs	BLMs
Unit 28 Decimals	**Number and Algebra** *Fractions and decimals* MA2-7NA represents, models and compares commonly used fractions and decimals Model and represent unit fractions including $\frac{1}{2}, \frac{1}{4}, \frac{1}{3}, \frac{1}{5}$ and their multiples to a complete whole (ACMNA058) AC	3.6 Number Line 3.12 Stopwatch 3.25 Place-Value Mat with Tenths	BLM 73 Tenths BLM 74 Fraction–Decimal Cards BLM 75 Blank Bingo Boards BLM 76 Decimal Cards BLM 77 Place-Value Mat with Tenths BLM 78 Human Reaction Time
Unit 29 Quadrilaterals and Symmetry	**Measurement and Geometry** *Two-dimensional space* MA2-15MG manipulates, identifies and sketches two-dimensional shapes, including special quadrilaterals, and describes their features Identify symmetry in the environment (ACMMG066) AC (Extension of Year 2) Describe and draw two-dimensional shapes, with and without the use of digital technologies (ACMMG042) AC	3.26 2D Shapes	BLM 10 Grid Paper BLM 79 2D Shapes BLM 80 Is it Symmetrical? BLM 81 Alphabet Cards
Unit 30 Volume and Capacity	**Measurement and Geometry** *Volume and capacity* MA2-11MG measures, records, compares and estimates volumes and capacities using litres, millilitres and cubic centimetres Measure, order and compare objects using familiar metric units of length, mass and capacity (ACMMG061) AC	3.28 Multilink Blocks	BLM 83 Investigating Millilitres

Unit	AC Strand/NSW Mathematics K–10 Syllabus Substrand & Outcome/AC Content Description & Code	NTOs	BLMs
Unit 31 3D Objects	**Measurement and Geometry** *Three-dimensional space* MA2-14MG makes, compares, sketches and names three-dimensional objects, including prisms, pyramids, cylinders, cones and spheres, and describes their features Make models of three-dimensional objects and describe key features (ACMMG063) AC		BLM 84 3D Object Templates 1 BLM 85 3D Object Templates 2 BLM 86 Prisms and Pyramids
Unit 32 Money	**Number and Algebra** *Addition and subtraction (money)* MA2-5NA uses mental and written strategies for addition and subtraction involving two-, three-, four- and five-digit numbers Represent money values in multiple ways and count the change required for simple transactions to the nearest five cents (ACMNA059) AC	3.27 Australian and International Coins	BLM 22 Banknotes BLM 87 Coins BLM 88 Coin Challenges BLM 89 Supermarket Catalogue BLM 90 Canteen List

Nelson Maths: Australian Curriculum NSW Scope and Sequence across the Year Levels

Note: the Working Mathematically outcomes of *Communicating*, *Problem Solving* and *Reasoning* are integrated throughout the activities and tasks in the program.

NSW Mathematics K–10 Syllabus strand and substrand	Kindergarten	Year 1	Year 2	Year 3	Year 4	Year 5	Year 6
Number and Algebra *Whole numbers*	Unit 1 Numbers to 5 Unit 2 Counting to 5 Unit 3 Groups of Things Unit 5 More Counting Unit 6 Dot Patterns Unit 8 Numbers to 10 Unit 9 Counting with Numbers to 10 Unit 10 Ten Frames Unit 11 Counting and Comparing Groups Unit 13 Ordinal Number Unit 16 Understanding More About Numbers to 10 Unit 22 Numbers Beyond 10 Unit 24 More About Numbers to 20	Unit 1 Recognising Numbers to 20 Unit 2 Counting to 20 Unit 5 Modelling Numbers Unit 6 Number Lines Unit 8 Numbers Beyond 20 Unit 15 2-digit Numbers Unit 16 More About 2-digit Numbers Unit 20 Money Unit 21 More About Money Unit 28 Place Value	Unit 1 Counting Unit 2 Modelling Numbers Unit 4 Numbers up to 1000 Unit 17 Money	Unit 1 Numbers! Numbers! Numbers! Unit 2 Numbers to 10 000 Unit 3 More About Numbers to 10 000 Unit 8 Place Value Unit 9 More About Place Value Unit 32 Money	Unit 1 Odd and Even Numbers Unit 2 Numbers to Tens of Thousands Unit 3 Place Value Unit 31 Money	Unit 1 Place Value and BODMAS Unit 3 Factors and Multiples Unit 6 Estimation	Unit 1 Place Value and BODMAS Unit 3 Integers Unit 6 Prime Numbers Unit 7 Composite Numbers Unit 18 Problems with Positive and Negative Numbers
Addition and subtraction	Unit 17 Beginning Addition Unit 19 More About Addition Unit 27 Subtraction Unit 28 More About Subtraction Unit 30 Addition, Subtraction and Money	Unit 12 Developing Mental Strategies for Addition Unit 25 Subtraction Unit 26 More About Subtraction Unit 30 Addition and Subtraction	Unit 4 Numbers up Unit 5 Strategies for Addition Unit 6 More Strategies for Addition Unit 9 Solving Problems with Addition Unit 11 Strategies for Subtraction Unit 12 Subtraction Unit 15 More About Subtraction Unit 16 Addition and Subtraction	Unit 5 Mental Strategies for Addition Unit 6 Addition Unit 8 Place Value Unit 12 Mental Strategies for Subtraction Unit 13 Subtraction Unit 14 Connections Between Addition and Subtraction Unit 15 Solving Addition and Subtraction Problems	Unit 13 Addition and Subtraction Unit 31 Money	Unit 2 Addition and Subtraction Unit 6 Estimation Unit 31 Financial Plans	Unit 2 All Four Operations

NSW Mathematics K–10 Syllabus strand and substrand	Kindergarten	Year 1	Year 2	Year 3	Year 4	Year 5	Year 6
Number and Algebra *Multiplication and division*	Unit 20 Grouping and Sharing	Unit 3 Skip Counting Unit 14 Grouping and Sharing	Unit 23 Multiplication Unit 24 More About Multiplication Unit 27 Division Unit 28 More About Division	Unit 18 Mental Strategies for Multiplication Unit 19 More About Mental Strategies for Multiplication Unit 22 Multiplication Unit 24 Division Unit 25 More About Division	Unit 6 Number Sequences: 3s, 6s and 9s Unit 7 Number Sequences 4s, 8s and 7s Unit 10 Multiplication Facts (Times Tables) Unit 11 Multiplication Facts and Related Division Facts Unit 15 Multiplication and Division Strategies Unit 16 More Multiplication and Division Strategies	Unit 3 Factors and Multiples Unit 7 Multiplication of Large Numbers A Unit 10 Multiplication of Large Numbers B Unit 11 Division with Remainders Unit 15 Mental Strategies	Unit 2 All Four Operations
Fractions and decimals	Unit 23 Halves	Unit 13 Halves and Quarters	Unit 7 Fractions Unit 30 More About Fractions	Unit 26 Fractions Unit 27 More About Fractions Unit 28 Decimals	Unit 18 Decimals to 2 Decimal Places Unit 23 Equivalent Fractions Unit 24 Counting with Fractions Unit 27 Fractions and Decimals	Unit 13 Addition and Subtraction of Decimals Unit 17 Fractions Unit 18 Decimals to Three Places Unit 23 Addition of Fractions Unit 24 Subtraction of Fractions Unit 27 Fractions and Decimals	Unit 14 Addition and Subtraction of Decimals Unit 15 Multiplication of Decimals Unit 17 Fractions Unit 23 Addition of Fractions Unit 24 Subtraction of Fractions Unit 27 Fractions of a Quantity Unit 28 Fractions, Decimals and Percentages Unit 31 Percentage Discounts
Patterns and algebra	Unit 14 Patterns	Unit 18 Patterns Unit 19 Number Patterns	Unit 18 Number Patterns Unit 19 More Number Patterns	Unit 21 Patterns	Unit 21 Number Patterns Unit 28 Number Sentences Unit 32 Word Problems	Unit 21 Number Patterns Unit 28 Number Sentences	Unit 11 Cartesian Systems Unit 21 Number Sequences

NSW Mathematics K–10 Syllabus strand and substrand	Kindergarten	Year 1	Year 2	Year 3	Year 4	Year 5	Year 6
Measurement and Geometry *Length* *Area*	Unit 7 Length and Area	Unit 10 Length and Area	Unit 3 Length Unit 25 Area	Unit 4 Length Unit 23 Area	Unit 4 Length and Temperature Unit 14 Perimeter and Area	Unit 4 Length, Area and Volume Unit 14 Perimeter Unit 16 Area	Unit 4 Area and Perimeter Unit 13 Length and Area Problems
Volume and capacity *Mass*	Unit 26 How Much Does It Hold? Unit 21 Mass	Unit 29 Capacity Unit 23 Mass	Unit 20 Capacity Unit 13 Mass	Unit 30 Capacity Unit 10 Mass	Unit 17 Volume Unit 5 Mass and Capacity	Unit 5 Mass and Capacity	Unit 16 Volume and Capacity Unit 5 Mass and Capacity
Time	Unit 15 Time Unit 29 More About Time	Unit 9 Time Unit 22 More AboutTime	Unit 14 Telling the Time Unit 21 More About Time	Unit 16 Time	Unit 25 Time Unit 26 Time Problems	Unit 25 Time Unit 26 Time Problems	Unit 25 Time Unit 26 Timetables Unit 12 Decimal Representations of the Metric System
Three-dimensional space	Unit 18 More About Shapes and Objects	Unit 17 3D Objects	Unit 26 3D Objects	Unit 31 3D Objects	Unit 9 Drawings of 3D Objects	Unit 9 3D Objects	Unit 8 2D Shapes and 3D Objects Unit 9 Prisms and Pyramids
Two-dimensional space	Unit 12 2D Shapes	Unit 4 2D Shapes	Unit 8 Transformation with 2D Shapes	Unit 29 Symmetry	Unit 8 Regular Shapes Unit 33 Patterns	Unit 8 Shapes Unit 32 Transformations	Unit 8 2 Shapes and 3D Objects Unit 32 Transformations Unit 33 Use of Transformations
Position	Unit 4 Position	Unit 7 Position	Unit 10 Position	Unit 7 Position	Unit 12 Mapping	Unit 12 Grid References	Unit 10 Mapping/Grid References Unit 11 Cartesian System
Angles				Unit 17 Angles	Unit 22 Angles	Unit 22 Angles Unit 33 Angle Applications	Unit 22 Angles
Statistics and Probability *Chance*		Unit 24 Chance	Unit 29 Chance	Unit 20 Chance	Unit 19 Chance	Unit 19 Chance	Unit 19 Chance
Data	Unit 25 Our Class	Unit 27 Data	Unit 22 Data	Unit 11 Our Community – Data	Unit 20 Collecting Data Unit 29 Displaying Data Unit 30 Interpreting Data	Unit 20 Collecting Data Unit 29 Data Displays Unit 30 Interpreting Data	Unit 20 Collecting Data Unit 29 Data Displays Unit 30 Interpreting Data

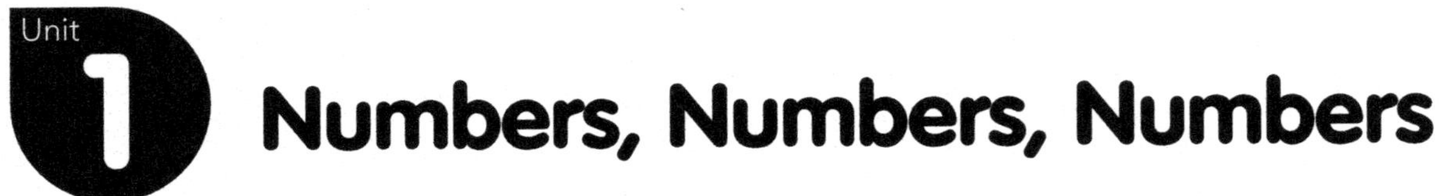

Unit 1 Numbers, Numbers, Numbers

Number and Algebra
Whole numbers MA2-4NA applies place value to order, read and represent numbers of up to five digits

even, hundreds, larger, odd, order, place value, smaller, tens, thousands, units

LESSON PLAN 1

TUNING IN

WHAT'S IN THE JAR?

You will need: beans, four jars or transparent containers

Present students with four jars of beans, with one jar containing 9 beans, the next jar containing 31 beans, the next 53 beans and the final jar 147 beans. Ask, 'Which jar has about 50 beans?' Have students discuss their guess and the reasons for their guess. Invite four students to count the beans in each jar. Record the numbers and tell students that these numbers have something in common that you will ask them about at the end of the lesson.

WHOLE-CLASS INTRODUCTION

EXPLORING ODD AND EVEN NUMBERS

You will need: sets of number cards made from BLM 1 '2-Digit Number Cards', counters

Have students work with a partner. Give each pair some of the number cards made from BLM 1 '2-Digit Number Cards'. Have students count out counters according to their number cards. Explain to students that they are going to determine if they have an odd or even number of counters and that this can be done by grouping counters in twos. If all the counters can be made into groups of two, the number is even. If there is a counter that cannot be grouped with another, the number is odd. Have students determine if their number is odd or even and make a list of the even numbers and the odd numbers.

INDEPENDENT TASKS

Note: Choose from Tasks 1, 2 or 3.

You will need: sticky dots, BLM 1 '2-Digit Number Cards', NTO 3.1 'Ten Frames', Student Book p. 4 'Odd or Even?'

TASK 1: DOT PAIRS

Have students write some 1- and 2-digit numbers, e.g. how old they will be next birthday, the last two digits in their phone number, their street number, the number of students in their class, etc. Students can then make models of the numbers by sticking dots (or drawing dots) in pairs on a sheet of paper and writing the number beside their model. Have students share and discuss their numbers with a partner and then label their models either odd or even.

TASK 2: INTERACTIVE TASK

Give students the number cards from BLM 1 '2-Digit Number Cards' that were not used in the 'Exploring Odd and Even Numbers' activity in the Whole-Class Introduction. Have them use NTO 3.1 'Ten Frames' to determine if the numbers are odd or even.

TASK 3: STUDENT BOOK p. 4 *'Odd or Even?'*

TEACHING GROUP

You will need: Unifix blocks, BLM 3 'Blank Chart'

FINGERS ON SHOW

- For students who require support, give them some Unifix blocks and have them make the numbers to 10 with two groups of blocks joined, e.g. 4 would have two groups of two blocks. Have students write a list of even numbers and odd numbers. Then have students play a game with a partner whereby they put one hand behind their back and decide how many fingers to hold up. On the count of three, both students show their hands. One student will score a point if the total number of fingers shown is even and the other will score a point if it is odd.

HOW DO YOU KNOW IT IS AN EVEN NUMBER?

- For students who require a challenge, give them a copy of BLM 3 'Blank Chart' and have them choose any starting number and fill in the numbers on the chart. Discuss with students the conditions for a number to be even and have them colour all of the even numbers. Have students look at the even numbers and write a rule for how to identify an even number.

REFLECTION

Select from the following to suit your class and their learning outcomes:

- Have students share their work from the Whole-Class Introduction or from the Teaching Groups, and ask, 'How do you know if a number is an even number?' Then invite students to name even numbers that are more than 20, 50, 100, 200, 500 and 1 000. Ask, 'Can you think of a rule for deciding if a number is odd or even?'
- Draw students' attention to the four numbers that you recorded during 'What's in the Jar?' in Tuning In, and ask them what the numbers have in common. Ask, 'How do you know that they are all odd numbers?'

LESSON PLAN 2

TUNING IN

TELL ME ABOUT THE NUMBER

Write a number on the board, e.g. 46, and ask students to think of as many things as they can about the number, e.g. it is an even number, it is made up of 4 tens and 6 ones, it is four less than 50, it is ten more than 36 and so on. Repeat for other numbers.

WHOLE-CLASS INTRODUCTION

WORDS AND NUMERALS

You will need: NTO 3.2 'Extended Notation: Number Cards', small whiteboards

Present NTO 3.2 'Extended Notation: Number Cards', selecting the random number to be shown. Have students read the number then invite a student to show the number with the number cards. Explain that we can see the number written as a numeral and shown with number cards. Ask, 'How can you write the number in words?' Have students say the number and then ask them to write the number in words on their whiteboards. To support students, write the words 'thousand', 'hundred', 'ninety', 'eighty', 'seventy', 'sixty', 'fifty', 'forty', 'thirty' and 'twenty' on the board or on cards. Repeat for other numbers.

INDEPENDENT TASKS

Note: Choose from Tasks 1, 2 or 3.

You will need: BLM 4 'Blank Cards', NTO 3.3 'Six-Sided Dice', Student Book p. 5 'Number Sort'

TASK 1: CARD GAME

Give each student two copies of BLM 4 'Blank Cards'. Have them make pairs of cards with numerals on half of the cards and the matching numerals in words on the other half. Students combine with a partner to play 'Memory', whereby all of the cards are placed face-down and students take it in turns to turn over two cards to find a pair. Alternatively, students could play 'Go Fish' or 'Snap'.

TASK 2: INTERACTIVE TASK 

Have students work with a partner using NTO 3.3 'Six-Sided Dice'. Before rolling the dice, students decide if they will form an odd or even number. They roll four dice and rearrange them to form a 4-digit odd or even number, which they record using words. Each time they meet the requirement, they score a point.

TASK 3: STUDENT BOOK p. 5 ***'Number Sort'***

TEACHING GROUP

You will need: NTO 3.3 'Six-Sided Dice', five dice for each student in the group

STEP BY STEP

- For students who require support, begin with 2-digit numbers and build up to 4-digit numbers. Present NTO 3.3 'Six-Sided Dice' and select two dice to randomly generate two digits. Have students write the two possible 2-digit numbers and determine if the numbers are odd or even. Have students write the numbers in words. Have students add 100 to the numeric form of their numbers and ask them to read the numbers and write them in words. Then have students add 1 000 to the numeric form of their numbers and have them read the numbers and write them in words. Repeat the process by rolling the two dice again.

FOUR CATEGORIES

- For students who require a challenge, have them work with higher numbers. Give each student five dice that they use to form four 5-digit numbers: the highest odd number, the highest even number, the lowest odd number and the lowest even number. Have students write their numbers in words and compare with the group. A point is given to the student who is able to form the highest or lowest in each category, and a point deducted if they are unable to form a number for any of the categories. Have students continue to roll the dice and form more numbers.

REFLECTION

Select from the following to suit your class and their learning outcomes:

- Have students discuss any difficulties they encountered when writing numbers in words and ask, 'How did you know that you had written the number correctly?'
- Create a chart of students' strategies for writing numbers in words. Display the chart in the classroom.

LESSON PLAN 3

TUNING IN

POPCORN!

Explain to students that they are going to play a game in which you will call out numbers, and if the number is odd they crouch down, but if the number is even, they jump up and say: 'Popcorn'. Have students stand, and begin by calling out 1- and 2-digit numbers. As the game progresses, increase the numbers to 3- or 4-digit numbers.

WHOLE-CLASS INTRODUCTION

MOVING NUMBERS

You will need: number cards made from BLM 2 'Mixed Number Cards'

Have students sit on chairs in a circle and give each student a number card made from BLM 2 'Mixed Number Cards'. Explain to students that, if their number meets the criteria mentioned, they need to stand up and swap chairs with someone else. Begin by giving criteria such as 'more than 500', 'odd number', '7 in the tens place', '4 units' or '4-digit number'. When students have had practice, say 'even number' and take away a chair as students are moving around so that one student is left in the centre of the circle. That student must give some criteria and try to find an empty chair to sit on while the others are moving.

INDEPENDENT TASKS

Note: Choose from Tasks 1, 2 or 3.

You will need: NTO 3.4 'Playing Cards', Student Book p. 6 'True Statement'

TASK 1: BEAT THE TEACHER

Model the following game, then have groups of three students play the game using NTO 3.4 'Playing Cards' on individual computers. One student must act as the teacher, while the other two students make the 4-digit number. In pairs, have students draw four boxes joined together in a horizontal line. Explain to students that the following activity involves them making the largest 4-digit number they can and both students need to agree where to place the numbers. Using NTO 3.4 'Playing Cards', generate a card and students decide in which box to place the number. (You must also choose a 4-digit number, but do not reveal it to the class.) The game continues until four cards have been generated and students have placed the numbers into boxes. Say your 4-digit number to the class. If a pair's number is lower than your number, they score 1 point, and if it is equal to your number, they score 3 points, but if it is higher than your number, they score 5 points. You score 10 points if you beat all the students. Note: when students are familiar with the game, make it more challenging by deciding that the number must be odd or even.

TASK 2: INTERACTIVE TASK

Have students work with a partner. Using NTO 3.4 'Playing Cards', the first student draws four cards that they rearrange to make the largest odd or even number they can. Their partner then draws four cards and tries to make a larger odd or even number.

TASK 3: STUDENT BOOK p. 6 *'True Statement'*

TEACHING GROUP

You will need: dice

BONGEL

- For students who require support, have them work with smaller numbers. Have students work with two or three dice, playing with a partner or in a small group. Each student rolls the dice to form the largest number they can. The student with the largest number in this round writes the letter 'B'. Students roll the dice again to form the largest number they can, and each time a student wins a round, they write a letter. The first student to form the word 'BONGEL' wins.

HOW MANY NUMBERS?

- For students who require a challenge, have them work with larger numbers. Give students five dice to roll and have them form and record the smallest number they can. Students roll the dice again to form a 5-digit number that is more than the previous number. Students continue to roll the dice, and each time they must form and record a number that is larger than the one before; if they cannot, they must stop. Have students compare how many numbers they formed.

REFLECTION

Select from the following to suit your class and their learning outcomes:

- Have students share the strategies they used to help them win the games they played.
- Draw four boxes on the board and tell students that you want them to form the largest number they can. Roll a 10-sided dice, and ask, 'Which box would you put the number in? Why do you think that is a good choice?'

Home Task

- Have students look at the newspaper (hard copy or online) for examples of numbers used in the media. Have them find examples of 4-digit numbers and what they are used for in daily life.

Assessment

- Have students complete **Student Assessment p. 7**.
- Review with students **Assessment Task Card 3.1**.

During the three lessons:

- Observe which students are able to read and write numbers in word and numeric form during 'Card Game' in Lesson Plan 2, Independent Tasks, Task 1, and mark on a class list.
- Make note of students completing the scaffolding tasks or the more challenging activities of the Teaching Groups.
- Review Student Book pages and make notes of areas of difficulty.

Recommendations for Future Learning

Specific to Student Assessment p. 7; if the student is experiencing difficulty:

Q 1 Have the student use counters to explore which numbers can make groups of two without any leftover counters. Have the student mark these even numbers on a 100 chart and identify patterns associated with even numbers.

Q 2–3 Include numbers in word form in weekly spelling lists.

Q 4 Have the student revisit what makes a number even or odd and have them model 2-, 3- and 4-digit numbers using MAB. Have them order numbers by comparing models.

If the student has not achieved the recommended skills for this unit:

1. See **Assessment Task Card 3.1** for specific recommendations.
2. Have the student work with 2- and 3-digit numbers in any of the listed activities or tasks prior to moving to 4-digit numbers.
3. Review *Nelson Maths: Australian Curriculum NSW Year 2* Unit 1.

If the student has achieved the recommended skills and these skills are firmly established, consider:

1. Having the student complete *Nelson Maths Building Mental Strategies Skill Book Year 4*, pp. 10–11, to reinforce mental strategies with 4- and 5-digit numbers.
2. Moving forward to *Nelson Maths: Australian Curriculum NSW Year 4* Unit 1.
3. Extending the student in any of the listed activities or tasks by using larger numbers.

Unit 2

Numbers to 10000

Number and Algebra
Whole numbers MA2-4NA applies place value to order, read and represent numbers of up to five digits

 digit, hundreds, MAB, number line, scale, tens, thousands

LESSON PLAN 1

TUNING IN

THE THOUSAND BLOCK

You will need: an MAB thousand, ten MAB hundreds

Hold up the MAB thousand and ask, 'Who has seen this before? What is it?' Discuss. Hold up an MAB hundred and ask, 'What is this? What is its value?' Discuss. Ask, 'How many hundreds would I need to make one thousand?' Hold ten MAB hundreds together to show that they form the same size as the MAB thousand. Reinforce that 10 hundreds is equal to 1 thousand.

WHOLE-CLASS INTRODUCTION

INTRODUCING ... THOUSANDS!

You will need: BLM 5 'Digit Cards', MAB (preferably with magnets)

Select four cards from BLM 5 'Digit Cards'. Stick them face down on the board, arranged as a 4-digit number. Turn over the card in the ones place and ask, 'What number is this?' Invite one student to represent this number with MAB. Turn the next card so that tens and ones are showing. Ask, 'What number is this?' Invite another student to add the tens to the MAB model. Repeat for the cards in the hundreds, then the thousands place so that a 4-digit number is identified and modelled. Repeat with another 4-digit number.

INDEPENDENT TASKS

Note: Choose from Tasks 1, 2 or 3.

You will need: decks of playing cards with 10s, Js, Qs and Ks removed; MAB class set; NTO 3.5 'Place-Value Dice'; Student Book p. 8: '4-digit Numbers'

TASK 1: PICK A WINNER

Provide pairs of students with a set of playing cards (10s, Js, Qs and Ks removed) and MAB. Shuffle the cards and place them between the students. Player A takes the first card from the top of the pile – this will be their ones value to model with MAB. Player B does the same. Player A selects the next card (tens) and so on until both players have a 4-digit number modelled with MAB. The player with the largest number scores one point; the winner is the first to score five points.

TASK 2: INTERACTIVE TASK 

Have students work in pairs on a computer, using NTO 3.5 'Place-Value Dice'. Have students roll the dice to create a 4-digit number and write this in the middle of a blank page. They roll again and decide whether the number is bigger or smaller than the original number. If smaller, they record it above the original number, if bigger, they record it below. Pairs continue until they have ten numbers.

TASK 3: STUDENT BOOK p. 8 *'4-digit Numbers'*

TEACHING GROUP

You will need: dice (enough for three or five per pair); MAB; decks of playing cards with 10s, Js, Qs and Ks removed

GOOD THINGS COME IN SMALL PACKAGES

- For students who require support, work with 3-digit numbers until they are able to extend understanding into the thousands. Have students work in pairs, taking turns to roll three dice, recording the number and modelling it with MAB. The student whose number is the smallest takes the lead. The student with the smallest number overall wins.

CREATING AND ORDERING 5-DIGIT NUMBERS

- For students who require a challenge, ask, 'What comes after thousands?' Lead discussion to the next value – tens of thousands – being ten times the value of the previous column. Ask, 'What would an MAB ten thousand look like?' Have students construct this using MAB thousands. Then, have them play 'Pick a Winner' from Independent Tasks, Task 1, using 5-digit numbers. Have students record numbers created, and after five games, order these from biggest to smallest.

REFLECTION

Select from the following to suit your class and their learning outcomes:

- Ask, 'How does modelling with MAB help you understand the value of a number?'
- Have students discuss their work from the Independent Tasks. Ask, 'How do you know if a number is bigger than another? Which number do you look at first?'

LESSON PLAN 2

TUNING IN

WHAT'S MY NUMBER?

Divide the class into four teams and have them sit in lines in front of the board. Write a number on scrap paper and keep it hidden from students. Draw four horizontal lines on the board to represent a 4-digit number. Say, 'I am thinking of a number in the thousands. Can you guess what it is?' The first member from Team A guesses a number by saying it in words, e.g. 'Two thousand, one hundred and thirty-six.' Write the numerals on the lines. Below the numerals, put a tick for correct place and digit, a dot for correct digit but incorrect place and a cross if the digit is not in the number. Team B then guesses. Write the numerals below the last guess. Fill in ticks, dots and crosses, and move to the next group. The group to discover the correct number wins.

WHOLE-CLASS INTRODUCTION

AN AUSSIE ADVENTURE

You will need: map of Australia or atlas, NTO 3.6 'Number Line'

Ask, 'If you could holiday anywhere in Australia, where would you go?' Record students' responses. Select a destination from the list, and say, 'I wonder how far away _____ is'. Use the map to show students their location and the destination. Allow students to guess how far away the destination is. Use an interactive map tool on the IWB to find the actual distance in kilometres. Using NTO 3.6 'Number Line' on the IWB, add this number in the centre of the blank number line. Select another location from the list, show its location on the map, and ask, 'Do you think this is closer or further away?' Find the distance and ask, 'Where does this number belong on the number line? Is it bigger or smaller than our first number?' If bigger, place it to the right of the number, if smaller, to the left. Repeat with two other locations. Note: this lesson focuses on order; scale is not important.

INDEPENDENT TASKS

Note: Choose from Tasks 1, 2 or 3.

You will need: BLM 2 'Mixed Number Cards', atlases, access to the internet, NTO 3.6 'Number Line', Student Book p. 9 'Road Trip!'

TASK 1: NUMBER CARDS

Have pairs of students use cards created from BLM 2 'Mixed Number Cards'. Put the cards upside down in a pile between the students. Students take turns to take a card and place it in order to form a number line. Encourage students to discuss and give reasons for their placement to the left (smaller) or right (larger). Once finished, students record the number line in their workbooks.

TASK 2: INTERACTIVE TASK 

Have students use an atlas to select six Australian destinations they would like to visit and record them in their workbooks. Using an interactive map tool, students find and record the driving distances from their location to their destinations. Have students order destinations from closest to furthest away. Students may put the numbers in order on NTO 3.6 'Number Line'.

TASK 3: STUDENT BOOK p. 9 *'Road Trip!'*

TEACHING GROUP

You will need: atlases, access to the internet, strips of paper

A STATE OF MY OWN

- For students who require support, work together to select four destinations in your state and use an interactive map tool on the IWB to work out the distance between their location and the destinations.

Have students write these on strips of paper and order the destinations from closest to furthest away. Students record their results in their workbooks.

THE WORLD AWAITS!

- For students who require a challenge, have them select six destinations from anywhere in the world and use the internet to find the distances from their location. Have students order the distances from largest to smallest.

REFLECTION

Select from the following to suit your class and their learning outcomes:

- Have students share their work from the Independent Tasks or Teaching Groups, and ask, 'How did you know whether a number was bigger or smaller? When was it difficult to put the numbers in order?'
- Play 'What's my Number?' from Tuning In, extending the number into the tens of thousands.

LESSON PLAN 3

TUNING IN

NEWSPAPER SCAVENGER HUNT!

You will need: newspapers, strips of paper, magnets

Give pairs of students one or two pages from a newspaper and one strip of paper. Ask, 'Where might you find 4-digit numbers in a newspaper?' (e.g. car sales, dates) Have students find a 4-digit number in their pages and write this down on their strip of paper. Invite one student from each pair to the front of the class, and examine their numbers. Have the students read their numbers aloud, then order themselves from smallest to largest. Tell remaining students to put their hands on their heads if they believe the students at the front are in the right order, or on the floor if they are wrong. Discuss. When the students are in the correct order, use magnets to attach the strips of paper in order on the board.

WHOLE-CLASS INTRODUCTION

PEGS FOR SCALE!

You will need: strips of paper from Tuning In with numbers written on them, a long piece of rope or string, pegs

Refer to the numbers on the board from the Tuning In activity. Roll out the rope and have two students hold it taut. Peg the smallest number at the far left of the rope. Say, 'We are using the rope as a number line, starting with our smallest numbers to the left, and getting bigger as we move to the right. We will examine how much bigger or smaller the numbers are and show this by putting more or less distance between them.' Take the second number from the board and ask, 'Is this number a lot or a little bigger than our first number?' Discuss and peg the number accordingly. Repeat with remaining numbers, selecting students to peg numbers in place. Explain that this process of varying the distance between numbers depending on their size is called using scale. Place an empty peg somewhere on the number line, and ask, 'What number could this be?' Discuss.

INDEPENDENT TASKS

Note: Choose from Tasks 1, 2 or 3.

You will need: newspapers, A3 paper, LO: *L2005 'Scale matters: hundreds'*, Student Book p. 10 'Line Them Up!'

TASK 1: NEWSPAPER NUMBER LINES

Have pairs of students search newspapers to find at least ten 4-digit numbers with a range of 1000, e.g. between 2000 and 3000. Have students rule a line across the width of a landscape sheet of A3 paper and form a number line by placing 11 equally spaced markers along the length of the line (counting by hundreds). Have students place their newspaper numbers to scale on the line. Ask, 'Are there any areas where you have lots of numbers? Or very few numbers? Are there any reasons for this?'

TASK 2: INTERACTIVE TASK 

Have students work independently on computers to play LO: *L2005 'Scale matters: hundreds'*, examining the given scale and filling in the appropriate value. This can also be played as a whole-class activity.

TASK 3: STUDENT BOOK p. 10 ***'Line Them Up!'***

TEACHING GROUP

You will need: newspapers, sticky notes, real-estate magazines, a long sheet of paper

SMALLEST TO LARGEST

- For students who require support, simplify by examining 3-digit numbers. Have students search the newspaper for 3-digit numbers and write these on sticky notes, until the group has about 20 numbers. Work as a group first to order the numbers from smallest to largest, and then to create a number line. Provide support and guidance as required.

HOUSES FOR SALE

- For students who require a challenge, give them a real-estate magazine where the prices are mostly 6-digit numbers. Have students work in pairs to select 20 houses, then to create a number line on a long sheet of paper to fit most (or all) of their house prices. Tell students to think about the scale they will need to use on their line. Discuss.

REFLECTION

Select from the following to suit your class and their learning outcomes:

- Have students present their number lines. Ask, 'Why is it important to scale our number lines rather than just write the numbers in order with equal spaces?'
- Have students each select one number from the newspaper activities in Tuning In or Independent Tasks, Task 1, write it on a piece of coloured card, then form a number line across the room. You could collect these to create a number line display in the classroom.

Home Tasks

Select from the possible Home Tasks:

- Have students look for things around the home that show numbers up to tens of thousands. This may include books, newspapers and items from the pantry. Have students list the items in categories of hundreds, thousands and tens of thousands.
- Have students teach a family member to play 'What's My Number?' from Lesson Plan 2, Tuning In, and extend the game to tens of thousands.

Assessment

- Have students complete **Student Assessment p. 11**.
- Review with students **Assessment Task Card 3.2**.

During the three lessons:

- Collect created items such as the newspaper number lines from, Lesson Plan 3, Independent Tasks, Task 1, to add to students' portfolios.
- Make note of students who completed the scaffolding tasks or the more challenging activities of the Teaching Groups.
- Review Student Book pages and make notes of areas of difficulty.

Recommendations for Future Learning

Specific to Student Assessment p. 11; if the student is experiencing some difficulty:

Q 1–2 Have the student model 3-digit numbers with MAB before moving on to 4-digit numbers.

Q 3–4 Revisit ordering activities from Lesson Plan 2. Have the student order and select the largest numbers from a set of 3-digit numbers. This could be scaffolded by using other concrete materials such as craft sticks.

Q 5 Revisit the 'Pegs for Scale!' activity from the Whole-Class Introduction in Lesson Plan 3. Revise the process of ordering numbers from smallest to largest, then move on to creating a number line to scale (perhaps using numbers on sticky notes to allow movement).

If the student has not achieved the recommended skills for this unit:

1. See **Assessment Task Card 3.2** for specific recommendations.
2. Have the student work with 2- and 3-digit numbers before moving on to larger numbers. Scaffold the student with the use of MAB or other concrete materials to reinforce the value of numbers.
3. Review *Nelson Maths: Australian Curriculum NSW Year 2* Unit 4.

If the student has achieved the recommended skills and these skills are firmly established, consider:

1. Having the student complete *Nelson Maths Building Mental Strategies Skill Book Year 3*, pp. 14–15, to reinforce mental strategies with ordering numbers.
2. Moving forward to *Nelson Maths: Australian Curriculum NSW Year 4* Unit 2.
3. Extending the student in any of the listed activities by using larger numbers, moving into hundreds of thousands and millions.

Unit 3

More About Numbers to 10 000

Number and Algebra
Whole numbers MA2-4NA applies place value to order, read and represent numbers of up to five digits

add, hundreds, MAB, ones, place value, place-value mat, spike abacus, subtract, tens, thousands

LESSON PLAN 1

TUNING IN

GUESS MY NUMBER

You will need: two chairs in front of the board, small whiteboard

Select two students to sit on the chairs with their backs to the board. Write a 4-digit number on the board. The students need to work out the number by taking turns to ask the class questions about it – the class may only answer 'yes' or 'no'. Encourage questions such as, 'Does the number have four digits? Is there a 7 in the number?' Record responses on a small whiteboard to aid thinking. Once questions are exhausted, the students may guess the number; the winner is the first to guess correctly. Select two new students to play again.

WHOLE-CLASS INTRODUCTION

ADDING AND SUBTRACTING 10, 100 AND 1 000

You will need: BLM 6 '4-Digit Number Cards', NTO 3.7 'Modelling with MAB'

Select a number from BLM 6 '4-Digit Number Cards'. Using NTO 3.7 'Modelling with MAB', invite a student to model the number (e.g. 4 519) with MAB, explaining how many MAB thousands they are selecting and why, how many hundreds and so on. Record the number below the model. Select another student to model the number that is 10 more than the previous number, and to name and record it. Compare the two numbers and ask, 'What is different? Why?' The digit in the tens place has increased by 1 because the number is 10 more. Continue modelling and recording numbers that are 10 more, 10 less; 100 more, 100 less; and 1 000 more and 1 000 less. For each, look at how the model and the written form differ.

INDEPENDENT TASKS

Note: Choose from Tasks 1, 2 or 3.

You will need: BLM 6 '4-Digit Number Cards', MAB class set or BLM 7 'MAB', BLM 8 'Place-Value Mat', NTO 3.7 'Modelling with MAB', Student Book p. 12 'Adding and Subtracting Value'

TASK 1: MODELLING 4-DIGIT NUMBERS

Give students four random numbers from BLM 6 '4-Digit Number Cards'. Have students order them from smallest to largest. Then, for each number, students use MAB or BLM 7 'MAB' to model the numbers that are 10, 100 and 1 000 more than these numbers. Students record their numbers on BLM 8 'Place-Value Mat'.

TASK 2: INTERACTIVE TASK 

Have students work independently on computers using NTO 3.7 'Modelling with MAB'. Give students a number from BLM 6 '4-Digit Number Cards' and have them model it on their screen. Students then make changes to the number: 10 more, 10 less; 100 more, 100 less; 1 000 more; 1 000 less. They model each new number on their screen and record it in their workbooks.

TASK 3: STUDENT BOOK p. 12 *'Adding and Subtracting Value'*

TEACHING GROUP

You will need: MAB, BLM 8 'Place-Value Mat', dice

EXPLORING MAB

- For students who require support, examine and compare MAB. Ask students to line up 10 ones to make 1 ten; 10 tens to make 1 hundred and so on. Then have students model 1 322 with MAB on BLM 8 'Place-Value Mat'. Have them add an MAB one to the number and ask, 'What number is this now? How has it changed? How would we write this new number?' Repeat this process, adding an MAB ten, hundred and thousand to the number. Repeat using subtraction.

REACHING HIGHER PLACES

- For students who require a challenge, have them work in pairs with each player beginning with the 5-digit number 11 111. Each player takes turns to roll a dice and assign any place value to the number they roll, e.g. a 3 could become 3, 30, 300, 3000 or 30000. They then add or subtract that number to 11 111. They continue until someone reaches 55555. Repeat the game with a different target number, e.g. 88888.

REFLECTION

Select from the following to suit your class and their learning outcomes:

- Write the numbers 2476 and 2486 on the board and ask, 'How are these two numbers different?'
- Ask, 'Which place value changes the value of a number the most? Which one changes it the least?'

LESSON PLAN 2

TUNING IN

IT'S MY BIRTHDAY!

You will need: sticky notes

Have students write the day and month of their birthday as a 4-digit number on a sticky note, e.g. September 21 would be 2109. Students then order themselves from smallest to largest – without talking, and by changing position only with the person next to them. Students read out numbers to check the order.

WHOLE-CLASS INTRODUCTION

THE IMPORTANCE OF ZERO

You will need: NTO 3.7 'Modelling with MAB', MAB

Display NTO 3.7 'Modelling with MAB' on the IWB and select a birthday number containing a zero in the tens column from Tuning In to model with MAB. Ask, 'What number is this? How would I write this number on the mat?' Discuss the importance of '0' and what this means – that there are no ones, tens etc. The zero holds the place and maintains the value of the other digits in the number. Look at the modelled number and ask, 'What number would be 10 less than this?' Have students explain their predictions. Invite a student to model the number. If students have difficulties, remind them that 10 tens are equivalent to 1 hundred, e.g. for 10 less than 2109, you need to break up the hundreds into tens and take 1 ten away, so it becomes 2099.

INDEPENDENT TASKS

Note: Choose from Tasks 1, 2 or 3.

You will need: BLM 9 'Zero's the Go!', decks of playing cards (with picture cards, jokers and 10s removed), calculators, dice, LO: *L871 'Wishball challenge: whole numbers'*, Student Book p. 13 'Birthday Values'

TASK 1: ZERO'S THE GO!

Have students work in groups of four. Give each student BLM 9 'Zero's the Go!' and ask them to play the game.

TASK 2: INTERACTIVE TASK 

Have students work independently on computers to play LO: *L871 'Wishball challenge: whole numbers'*, adding or subtracting numbers according to place value as they work towards a target number. They use the 'wishball' to select the final digit. This can also be played as a whole-class activity.

TASK 3: STUDENT BOOK p. 13 *'Birthday Values'*

TEACHING GROUP

You will need: craft sticks (about 20 single and at least 10 bundles of ten), BLM 8 'Place-Value Mat', MAB

MAKING AND BREAKING BUNDLES

- For students who require support, work with craft sticks on BLM 8 'Place-Value Mat' to reinforce that 10 ones equals 10, 10 tens equals 100 and so on. Place nine single sticks in the ones column on the mat and ask, 'What happens if I add one more stick to my ones?' Show that we now have ten, then bundle the sticks together and move them into the tens column. Repeat with tens to hundreds. Once the concept is understood using the sticks, demonstrate with MAB, and move their understanding into the thousands.

INTO THE MILLIONS!

- For students who require a challenge, have them add their abbreviated year of birth to their birthday number from Tuning In, e.g. if born in 2004, they add 04 to the end of their number. Have the group share their numbers. Discuss the place value for the first two digits in their numbers. Ask students to order the numbers from smallest to largest. Have students subtract 10, 100, 1000, 10000 and 100000 from their number.

REFLECTION

Select from the following to suit your class and their learning outcomes:

- Ask students to move back into the line from the Tuning In activity and name the number that is 10, 100 or 1 000 more or less than their birthday number.
- Ask, 'What is the importance of zero in a number?'

LESSON PLAN **3**

TUNING IN

THE SPIKE ABACUS

You will need: a spike abacus, NTO 3.8 'Spike Abacus'

Hold up a spike abacus and ask, 'Who has seen this before? What is it?' Discuss. Display NTO 3.8 'Spike Abacus' on the IWB with 'show thousands' selected. Select a 'random number' to be shown. Ask, 'What number is shown? How do you know?' Add beads to the ones column so there are nine beads. Ask, 'What will happen if I add one more?' Discuss. There cannot be more than 9 ones – adding one more will make a 'ten' bead, so all of the beads are taken off the ones spike, and added to the tens. Add the bead and watch the movement. Continue by looking at other numbers and adding beads to other spikes.

WHOLE-CLASS INTRODUCTION

CREATING NUMBERS

You will need: NTO 3.8 'Spike Abacus'

Write the numbers 0, 2, 4 and 6 on the board. Ask, 'How many numbers can you make using these digits?' Students list their numbers on a blank page. Ask, 'How many numbers did you make? Did anyone have a strategy?' Display NTO 3.8 'Spike Abacus' on the board with both abacuses displayed. Have one student choose a number from their list and write it in the box below one abacus. Have them model the number on the abacus. Reinforce the value of each number, e.g. a 4 on the tens spike is worth 40; a 2 on the thousands spike is worth 2000. Select another student to do the same using the other abacus on the IWB. Discuss differences between the numbers – while each uses the same digits, the order of the digits determines the value of the number.

INDEPENDENT TASKS

Note: Choose from Tasks 1, 2 or 3.

You will need: dice (four per pair), spike abacus (one for each student), NTO 3.8 'Spike Abacus', Student Book p. 14 'Numbers on an Abacus'

TASK 1: SPIKE ABACUS CHALLENGE

Have students work in pairs with four dice and a spike abacus each. Student A rolls the four dice, creates the largest number they can and models it on their abacus. Student B does the same. Each number will now undergo four changes. Student A rolls one dice, and decides which spike they will add this amount to; they repeat this three times, each time adding the number to a different spike. Student B has their turn. Students then compare the values on the two abacuses; the student with the largest number receives one point. The winner is the first to get to five points.

TASK 2: INTERACTIVE TASK

Have students work independently on computers using NTO 3.8 'Spike Abacus'. Give students four digits, e.g. 0, 3, 7, 9, and have them use these to create ten different numbers in their workbooks. Students model each number in turn on the abacus and answer the question, 'How does your number change if you add three more beads to the tens prong?' Students should add the beads and write a response in their workbooks.

TASK 3: STUDENT BOOK p. 14 ***'Numbers on an Abacus'***

TEACHING GROUP

You will need: dice (five per student), BLM 8 'Place-Value Mat', spike abacus

LOOKING AT 3-DIGIT NUMBERS

- For students who require support, give them three dice to roll a 3-digit number. They then use BLM 8 'Place-Value Mat' (folded to hide the thousands column) to write the largest number they can. Model the number on the spike abacus and discuss place-value components of the number, e.g. 'There are 2 hundreds, 3 tens and 7 ones.' Using the same digits, students create the smallest number possible, record it on the place-value mat and model it on the abacus. Repeat as required. Once students are confident, give them a fourth dice and have them create 4-digit numbers on the place-value mat (unfolded) and abacus.

CREATING AND ORDERING 5-DIGIT NUMBERS

- For students who require a challenge, given them five dice and ask them to draw a 5-digit place-value mat in their workbooks. Have them play the spike abacus challenge from Independent Tasks, Task 1, with the variation of starting with the largest number and using five single dice rolls to subtract beads from their abacus to reach the smallest number they can. Have them track their progress on their place-value mat.

REFLECTION

Select from the following to suit your class and their learning outcomes:

- Ask, 'How is the spike abacus different/similar to MAB? Do you find it easier or more difficult to understand? Why?'
- Have students share their results from the spike abacus challenge in Independent Tasks, Task 1, and ask, 'How did you decide what changes to make to reach the largest number?

Home Tasks

Select from the possible Home Tasks:

- Have students compile a list of their families' birthdays as 4-digit numbers, order them from largest to smallest and write them in words.
- Have students look at numbers around their home. Are there any numbers where a zero has no meaning? (e.g. the zeros before the numbers on the car odometer)

Assessment

- Have students complete **Student Assessment p. 15**.
- Review with students **Assessment Task Card 3.3**.

During the three lessons:

- Make note of how effectively students use concrete materials to model their numbers.
- Make note of students who completed the home tasks. Comment on areas of difficulty students may have experienced.
- Review Student Book pages and note areas of difficulty.

Recommendations for Future Learning

Specific to Student Assessment p. 15; if the student is experiencing some difficulty:

Q 1–2 Have the student use digit cards to manipulate the order of the digits to create the largest number. Model this with MAB and discuss the properties of the number.

Q 3–4 Use MAB on a place-value mat to model numbers, then add or subtract 10 or 100.

Q 5 Revise the 'Making and Breaking Bundles' activity from Lesson Plan 2, Teaching Group, to consolidate understanding of the relationships between values.

If the student has not achieved the recommended skills for this unit:

1. See **Assessment Task Card 3.3** for specific recommendations.
2. Have the student complete 'Add 10' and 'Take Away 10' from *Nelson Maths Building Mental Strategies Big Book 3*, pp. 4–5.
3. Have the student work with 2- and 3-digit numbers before moving on to larger numbers. Scaffold the student with the use of MAB, the spike abacus and other concrete materials to reinforce the value of numbers.
4. Review *Nelson Maths: Australian Curriculum NSW Year 2* Unit 2.

If the student has achieved the recommended skills and these skills are firmly established, consider:

1. Having the student complete *Building Mental Strategies Skill Book Year 3*, pp. 10–11, to reinforce understanding of place-value units.
2. Moving forward to *Nelson Maths: Australian Curriculum NSW Year 4* Unit 2.
3. Extending the student in any of the listed activities by using larger numbers, moving into hundreds of thousands.

Unit 4 Length

Measurement and Geometry

Length MA2-9MG measures, records, compares and estimates lengths, distances and perimeters in metres, centimetres and millimetres, and measures, compares and records temperatures

centimetre, cubits, estimate, foot, height, length, metre, millimetre, palm, ruler, shortest, tallest, tape measure, trundle wheel, width

LESSON PLAN 1

TUNING IN

FORMAL AND INFORMAL UNITS

Explain to students that in ancient times there were no standard measurements of length, such as centimetres and metres. One of the most common units was a cubit. A cubit is the length of the arm from the tip of the fingers to the elbow. Ask students to look at their 'cubit' and compare it with somebody else's. Discuss potential problems in using informal units and the advantages of using centimetres and metres as formal units of length.

WHOLE-CLASS INTRODUCTION

MEASURING IN CENTIMETRES

You will need: 30 cm rulers

Give each student a 30 cm ruler and ask, 'What do we use these for?' Promote discussion about centimetres and what we use them to measure. Have students share experiences in using centimetres. Explain that we use centimetres to measure length, height and width. Discuss these dimensions using an object such as a tissue box as a reference. Ask, 'How do I use my ruler to measure? Where do I start measuring from?' Discuss, emphasising the need to start measuring at zero, not at 1 or the very end of certain types of rulers.

INDEPENDENT TASKS

Note: Choose from Tasks 1, 2 or 3.

You will need: rulers, string, paper, *PowerPoint*, Student Book p. 16 'My Hand in Centimetres'

TASK 1: USING CUBITS TO ESTIMATE LENGTH

Have students compare the length of their cubit with a 30 cm ruler and use this to estimate something that is 30, 60 or 90 cm long. Have students check their estimates using a 30 cm ruler and compare their results with those of other students.

TASK 2: INTERACTIVE TASK 

Have students work in pairs to find classroom items that are the same length as their hand, their foot, their cubit and their finger. Have them take digital photos of the items next to their body part (e.g. their hand next to a whiteboard duster), then make a *PowerPoint* presentation of body measures with the photos.

TASK 3: STUDENT BOOK p. 16 *'My Hand in Centimetres'*

TEACHING GROUP

You will need: paper

ESTIMATING LENGTH USING THE HAND

- For students who require support, help them make a 10 cm cutout of their hand (see Student Book, page 16). Use the paper hand to estimate lengths in multiples of 10 cm. Check the estimates using a ten stick or a 30 cm ruler.

ESTIMATING LENGTH MORE ACCURATELY

- For students who require a challenge, have them think about what they could use to estimate lengths that are in between multiples of 10 cm. For example, a 10 cm hand span and two little finger widths might produce an accurate estimate for something measuring 12 cm.

REFLECTION

Select from the following to suit your class and their learning outcomes:

- Review Independent Tasks, Tasks 1 and 2. Ask, 'Did everyone get the same result?' Lead the discussion to consider why uniform units are necessary.
- Ask, 'What did you do when your whole foot, cubit, hand didn't fit exactly?' (Did they use half or quarter measures?) 'Would using smaller units solve this problem?'

LESSON PLAN 2

TUNING IN

WHAT IS A METRE?

You will need: masking tape, classroom measuring devices (30 cm rulers, metre ruler, tape measure)

Prior to the class, stick a 1 m piece of masking tape on the floor. Have students sit in a circle around the tape. Ask, 'How long is this piece of tape?' Have students estimate the length, then ask, 'How do you know?' to gain an idea of strategies used. Ask, 'What could we use to check the length?' Have students provide suggestions and measure the length of the tape. Students should realise that the line is exactly 1 m, and that 1 m is the same as 100 cm.

WHOLE-CLASS INTRODUCTION

MEASURING IN METRES

You will need: metre ruler, tape measure, trundle wheel

Ask, 'What kinds of things do we measure in metres?' Record students' responses on the board. Students should realise that we use metres to measure large objects or distances. Ask, 'How do we measure metres? What can we use?' Present the metre ruler, tape measure and trundle wheel to the class. Discuss each in turn – how we use them and what we might use them for. Record responses on the board. As each measuring device is examined, measure the metre line on the floor from the Tuning In activity to demonstrate how the device is used, focusing on where to begin measuring. This will show students that each device accurately measures the same distance.

INDEPENDENT TASKS

Note: Choose from Tasks 1, 2 or 3.

You will need: measuring equipment (e.g. metre rulers, tape measures, trundle wheels, string), LO: *L1075 'Direct a robot: how far?'*, Student Book p. 17 'How Long?'

TASK 1: MEASURING THE BASKETBALL COURT

Have small groups of students work out how they could measure the length of the basketball court (or hall, classroom etc.). Provide a range of measuring equipment and have students decide what they will use and how they will use it. Have students measure the distance, record their result and compare this with other groups.

TASK 2: INTERACTIVE TASK 

Have students work independently on computers to play LO: *L1075 'Direct a robot: how far?'*, deciding on the best distance to move the robot so that it can collect soil and rock samples from the Moon.

TASK 3: STUDENT BOOK p. 17 *'How Long?*

TEACHING GROUP

You will need: 30 cm rulers, objects to measure, access to the internet or library

MEASURING CLASSROOM ITEMS

- For students who require support, spend more time looking at centimetres and measuring with 30 cm rulers. Provide an assortment of objects for measuring, such as books, stapler, tissue box, and have students estimate and then measure the length of each object. Emphasise the need to start measuring at zero. Have students line the objects up in order of length.

IN OTHER COUNTRIES

- For students who require a challenge, have them research the imperial system of measurement. Students find out which countries use this system and record the imperial units. Have students compare the size of these units of measurement to metric units. If there is time, students could create a 30 cm ruler that has metric units on one side and imperial units on the other.

REFLECTION

Select from the following to suit your class and their learning outcomes:

- Have students who completed Student Book p. 17 'How Long?' discuss their results. Ask, 'Was it difficult to find objects that were exactly one metre?'
- Ask, 'Can you think of any real-life examples when you need to be able to measure length?'

LESSON PLAN 3

TUNING IN

LESS THAN A CENTIMETRE

Using the 'Think, Pair, Share' technique, ask students to visualise something that they think is about half a centimetre long. Have them share their idea with a partner and discuss whether one or both ideas are appropriate. Each pair then shares their ideas with another pair and this can lead to a group or class list being made. Repeat the activity with the challenge, 'What is something that is less than half a centimetre long?' Introduce the concept of a millimetre as a unit of length.

WHOLE-CLASS INTRODUCTION

MEASURING IN MILLIMETRES

Remind students of the millimetre as a unit of length for measuring in parts of centimetres. They may wish to imagine a millimetre as about the thickness of a finger nail. Ask students to think of a person who is likely, at this very moment, to be using millimetres in their work. Use the 'Think, Pair, Share' technique to promote discussion about why it is sometimes necessary to use millimetres when measuring length in the real world.

INDEPENDENT TASKS

Note: Choose from Tasks 1, 2 or 3.

You will need: 30 cm rulers with millimetre graduations, Student Book p. 18 'Smaller than a Centimetre'

TASK 1: MEASURING A PAPER CLIP

Show a paper clip that you have already measured and inform students that it is not an exact number of centimetres long. Have students estimate its length. Compare answers. Ask students to find the exact length using a ruler. Repeat for other objects if desired.

TASK 2: INTERACTIVE TASK 

Ask students to find the difference between the sizes of a $1 and $2 coin. They could use NTO 3.23 'Area of Shapes' for this purpose

TASK 3: STUDENT BOOK p. 18 *'Smaller than a Centimetre'*

TEACHING GROUP

You will need: 30 cm rulers with millimetre graduations, small objects to measure

MEASURING CLASSROOM ITEMS

- For students who require support, spend some time measuring accurately with a 30 cm ruler, ensuring that they place the ruler accurately when reading a measurement. Following this, read measurements using terms such as 'half a centimetre', 'a bit more than, less than a centimetre' and so on before measuring in millimetres.

LENGTH CONVERSION

- For students who require a challenge, have them convert between lengths expressed in millimetres and lengths expressed in centimetres and millimetres. They could also research the use of millimetres for longer lengths in the building industry, such as a wall height of 2400 mm.

REFLECTION

Select from the following to suit your class and their learning outcomes:

- Can students think of anything that is a millimetre long (or less)?
- Ask, 'In what sort of work is it important to be accurate to one millimetre when measuring length?'

Assessment

- Have students complete **Student Assessment p. 19**.
- Review with students **Assessment Task Card 3.4**.

During the three lessons:

- Make note of how students work with rulers, particularly whether they start ruling from zero.
- Collect the *PowerPoint* presentations created in Lesson Plan 1, Independent Tasks, Task 2, to add to students' digital portfolios.
- Throughout Lesson Plan 2, ask, 'What is length? How do we measure length?' and track responses.

Recommendations for Future Learning

Specific to Student Assessment p. 19; if the student is experiencing difficulty:

Q 1 Have the student look at rulers and tape measures to look for the unit used. Revise content from Lesson Plans 2 and 3.

Q 2 Remind the student of the types of informal measures used in Lesson Plan 1, particularly the body parts, and the problems associated with them.

Q 3 Revisit the activities on using rulers to measure centimetres in Lesson Plan 2. Remind the student that in measuring length, they need to start at zero.

Q 4 Remind the student of the meanings of the words 'shortest' and 'longest'. Revise ordering activities from Unit 2 'Numbers to 10 000'.

Q 5 Show the student real-life examples of these devices. Read the names to them if they are having difficulties.

If the student has not achieved the recommended skills for this unit:

1. See **Assessment Task Card 3.4** for specific recommendations.
2. Have the student revisit working with informal measures by comparing the length of a group of objects to identify which is the longest or shortest.
3. Have the student continue to work with concrete materials such as rulers or tape measures to increase familiarity. Have them liken these materials to number lines to build on prior knowledge.
4. Review *Nelson Maths: Australian Curriculum NSW Year 2* Unit 3.

If the student has achieved the recommended skills and these skills are firmly established, consider:

1. Having the student learn to convert between metres and centimetres.
2. Having the student extend knowledge of units of measurement to kilometres.
3. Moving forward to *Nelson Maths: Australian Curriculum NSW Year 4* Unit 4.

Unit 5

Mental Strategies for Addition

Number and Algebra

Addition and subtraction MA2-5NA uses mental and written strategies for addition and subtraction involving two-, three-, four- and five-digit numbers

add, adding 10s, build to 10, doubles, hundred chart, near doubles, number line, strategy, subtract

LESSON PLAN 1

TUNING IN

ADDING 10s

You will need: sticky notes, MAB

Have each student write their favourite 2-digit number on a sticky note and then model it with MAB. Write '+ 10' on the board and have students add 10 to their number. Encourage students to use MAB and emphasise that '+ 10' means adding another ten. Have students use the MAB to add 30, 50 or 100 to their number. Ask, 'How could this strategy help you with solving 63 + 24?' Talk about how students could identify and add the tens in the numbers and then add on the ones.

WHOLE-CLASS INTRODUCTION

JUMP ON THE LINE

You will need: masking tape or chalk

Mark a long line on the floor using masking tape or chalk. Tell students that it is an open number line. Present the problem: 37 + 20. Have a student demonstrate how they would use the number line to solve the problem. For example, imagining the start of the line is 37, they may add 10 and jump on the line to 47, then add another 10 and jump to 57. Encourage the student to explain how they use the adding 10 strategy. Repeat with other problems, e.g. 54 + 30 and 61 + 24. Ask, 'How would you solve 42 + 27 on the open number line?'

INDEPENDENT TASKS

Note: Choose from Tasks 1, 2 or 3.

You will need: BLM 12 'Sports Equipment Prices', LO: *L94 'The part-adder: generate hard sums',* Student Book p. 20 'Adding 10s'

TASK 1: SPORTS EQUIPMENT PRICES

Give students BLM 12 'Sports Equipment Prices'. Explain that the sports equipment store has different prices for each day of the week. All sports equipment is increased by $10 on Monday, $30 on Tuesday, $50 on Wednesday, $100 on Thursday and $300 on Friday. Have students calculate how much the following items cost on each day of the week: soccer ball, tennis racquet, hockey stick, basketball, running shoes, skipping rope.

TASK 2: INTERACTIVE TASK

Have students work with partners on computers, using LO: *L94 'The part-adder: generate hard sums'*. Encourage them to solve the addition problems by using the adding tens strategy. For example, for the equation 45 + 32, students can identify and add on the 3 tens in 32 and then add on the 2 ones. This means the equation would be extended to 45 + 30 + 2 = 77.

TASK 3: STUDENT BOOK p. 20 *'Adding 10s'*

TEACHING GROUP

You will need: MAB, NTO 3.9 '100 Chart', BLM 13 'Addition Cards'

ADDING TENS

- For students who require support, encourage them to use MAB to model the multiples of 10 up to 100 (10, 20, 30 etc.). Have them add together different multiples of 10. For example, ask, 'What does it make when 30 and 50 are added together?' Have students explain how they worked out the answers.

100 CHART

- For students who require a challenge, present NTO 3.9 '100 Chart' (with all numbers shown) and a set of cards made from BLM 13 'Addition Cards'. Have students take cards and then use the 100 chart to help them work out the answers. For example, for the equation 54 + 21, students would identify 54 as their starting point on the 100 chart. They would then move two spaces down to 74 (because 20 has been added) and then move one space to the right to 75 (because 1 has been added).

REFLECTION

Select from the following to suit your class and their learning outcomes:

- Have students share their work from Independent Tasks, Task 2. Ask, 'How did the adding ten strategy help you to work out these addition sums?'
- Encourage students to reflect on Independent Tasks, Task 1. Discuss the strategy they used to calculate the prices. Ask, 'What was the best day to go shopping at the store?'

LESSON PLAN 2

TUNING IN

BUILD TO 10 SNAP

You will need: BLM 5 'Digit Cards'

Have pairs of students play a game of 'Snap' with cards made from BLM 5 'Digit Cards'. Students can 'snap' when they match two numbers that go together to make 10, e.g. when a 4 and 6 are placed on top of each other.

WHOLE-CLASS INTRODUCTION

USING BUILD TO 10

You will need: NTO 3.10 'Card Flip', counters

Present students with NTO 3.10 'Card Flip', selecting cards with numbers from 0–99. As the numbers are shown, have students hold up fingers above their head to show how many more is needed to build to the next ten. For example, if 24 is shown, students would hold up six fingers to build to 30. Write the equation 24 + 8 on the board. Ask, 'How could we use the build to 10 strategy to help solve this problem?' Explore how the 8 is made up of 6 and 2, and when this 6 is with the 24 it would build to 30 and then the other 2 could be added to make 32 (i.e. 24 + 6 + 2 = 32). Use counters to model this if necessary.

INDEPENDENT TASKS

Note: Choose from Tasks 1, 2 or 3.

You will need: BLM 14 '100 Chart', dice, counters, small whiteboards, NTO 3.10 'Card Flip', BLM 5 'Digit Cards', Student Book p. 21 'Build to 10'

TASK 1: DICE RACE

Give pairs of students BLM 14 '100 Chart', dice, counters and a small whiteboard. To start the game, students place their counter on number 1 on the 100 chart. They take turns to roll a dice. The square they are positioned in and the number on the dice form an addition equation that they record on the whiteboard. For example, if the student is on square 25 and they roll a 6, they would record the equation 25 + 6. Encourage students to use the build-to-10 strategy when answering the problems, e.g. in this case it would be 25 + 5 + 1 = 31. The answer to the addition problem is the square the student moves their counter to. The winner is the first player to reach 100.

TASK 2: INTERACTIVE TASK

Present students with NTO 3.10 'Card Flip' (selecting 0–99 cards) and a set of cards made from BLM 5 'Digit Cards'. As they are shown the 2-digit number on the NTO, have students flip over one of the digit cards. They then make an addition equation with the two numbers, e.g. 46 + 7. Have students use the build-to-10 strategy to solve the problems.

TASK 3: STUDENT BOOK p. 21 *'Build to 10'*

TEACHING GROUP

You will need: Unifix blocks, LO: *L94 'The part-adder: generate hard sums'*

TOWERS OF 10

- For students who require support, revise the numbers that add together to make 10. Have them make towers using different coloured Unifix blocks to model the numbers that make 10, e.g. a tower with one red and nine blue blocks. Have students make a poster showing the combinations of numbers that make 10, e.g. 8 and 2. Students could then refer to their poster when they attempt building to 10 when solving equations such as 26 + 7.

MORE BUILDING

- For students who require a challenge, have them work with a partner using LO: *L94 'The part-adder: generate hard sums'*. Guide students as they explore how they could use the build-to-10 strategy when solving the 2-digit by 2-digit addition equations. Discuss other strategies, e.g. adding multiples of 10.

REFLECTION

Select from the following to suit your class and their learning outcomes:

- Present the equations: 8 + 6, 34 + 7, 158 + 3. Encourage students to explain how they would use the build-to-10 strategy in solving these equations. Ask, 'How does this strategy help you with solving addition problems?'
- Have students explain how knowing addition facts to 10 helps with using the build-to-10 strategy. Ask students to identify activities where they used this knowledge.

LESSON PLAN 3

TUNING IN

MAKING DOUBLES

You will need: counters, small whiteboards

Have students work in pairs to model doubles equations with counters and record the addition equations on a small whiteboard, e.g. 6 + 6 = 12. Ask, 'How might knowing doubles help you with solving addition problems?' Present the equation: 6 + 7. Have students explore and model how this equation is close to the doubles fact 6 + 6 and how they can use this to help them solve 6 + 7. Ask, 'Why would near doubles be a useful strategy?'

WHOLE-CLASS INTRODUCTION

FIND A DOUBLE AND NEAR DOUBLE

You will need: BLM 15 'Number Cards 1–20'

Make two sets of cards from BLM 15 'Number Cards 1–20'. Give each student a card and have them move around the room to find a person who is holding the card that is their double. Have students solve their double equation together, e.g. 6 + 6 = 12. Have students move around the room again, this time looking for someone with a card that is their near double. For example, a student holding 6 would need to find someone holding 7 or 5. Ask students to solve their near double equation, e.g. 6 + 7 = 13. Discuss the strategies they used to work out their answer. Have students swap cards and repeat the activity.

INDEPENDENT TASKS

Note: Choose from Tasks 1, 2 or 3.

You will need: BLM 16 'Near Doubles', NTO 3.10 'Card Flip', small whiteboards, Student Book p. 22 'Doubles and Near Doubles'

TASK 1: NEAR DOUBLES MEMORY

Give pairs of students a set of cards made from BLM 16 'Near Doubles'. Have them place the cards face down and play 'Memory', trying to find a near doubles equation and its answer. If the student successfully matches an equation and answer, they have another turn. The winner is the person who has the most pairs.

TASK 2: INTERACTIVE TASK 

Present students with NTO 3.10 'Card Flip', selecting 0–20 cards. As the numbers are shown, have students generate a doubles and a near doubles equation that uses that number. Students can record their equations on small whiteboards.

TASK 3: STUDENT BOOK p. 22 *'Doubles and Near Doubles'*

TEACHING GROUP

You will need: dominoes, BLM 17 'Blank Dominoes'

DOUBLES AND NEAR DOUBLES DOMINOES

- For students who require support, have them find dominoes that represent a doubles equation. For example, a domino with five dots and five dots would represent 5 + 5 = 10. Support students in recognising that doubles dominoes have the same amount on each end. Have students then find dominoes that represent near doubles equations. For example, a domino with six dots and seven dots would represent 6 + 7 = 13. Emphasise that the near doubles dominoes have an extra dot on one end.

BIGGER DOMINOES

- For students who require a challenge, give them BLM 17 'Blank Dominoes' and have them draw dots to make their own dominoes that represent doubles or near doubles equations, e.g. 10 dots and 11 dots. Have them swap dominoes with a partner and solve the problem. Encourage them to explain the strategies they used. Students could then make dominoes using numbers larger than 20, writing the number rather than drawing the dots.

REFLECTION

Select from the following to suit your class and their learning outcomes:

- Have students explain their understanding of the doubles and near doubles strategy. Allow students to use counters and a whiteboard if necessary. Ask, 'How would this strategy help you solve addition problems?'
- As a group, brainstorm and record known doubles and near doubles equations on a large sheet of paper. Display the chart in the room for students to refer to during future maths sessions.

Home Tasks

Select from the possible Home Tasks:

- Have students find numbers around the house, for example in magazines, on the remote control or in cookbooks. Encourage them to make addition equations using the numbers and then solve them by using mental addition strategies.
- Have students look at the numbers on letter boxes in their street and add 10, 30 or 100 to the numbers.

Assessment

- Have students complete **Student Assessment p. 23**.
- Review with students **Assessment Task Card 3.5**.

During the three lessons:

- During tasks, observe which students are able to solve addition equations by using the strategies of adding tens, building to 10 or near doubles.
- Make a note of students who completed the scaffolding tasks or the more challenging activities in the Teaching Groups.
- Review Student Book pages and make notes of areas of difficulty.

Recommendations for Future Learning

Specific to Student Assessment p. 23; if the student is experiencing some difficulty:

Q 1–2 Have the student review the connection between '+ 10' and adding a group of 10. Encourage the student to use MAB to model adding 10, 30 and 40 to 2-digit numbers. Emphasise how the tens increase when solving these addition problems.

Q 3 Revise the numbers that go together to make 10. Give a number between zero and 10 and have the student say the number that is needed to build to 10, for example, 4 and 6. Then give a 2-digit number and have the student build to the next ten, for example, 37 and 3.

Q 4 Revise doubles facts to 10. Encourage the student to model the doubles equations using Unifix blocks. Then connect this understanding to the near doubles strategy. Have students model a near doubles equation with blocks. Take away the 'extra' block so that it becomes a double. Emphasise how to use knowledge of doubles when solving near doubles.

Q 5 Revise the mental addition strategies that have been explored. Have the student make a poster of the strategies that they know how to use. Encourage the student to refer to the poster when solving problems.

If the student has not achieved the recommended skills for this unit:

1. See **Assessment Task 3.5** for specific recommendations.
2. Have the student revise strategies for solving simple addition problems, such as counting on and build to 10.
3. Provide more opportunities for the student to practise and consolidate specific mental addition strategies.
4. Review *Nelson Maths: Australian Curriculum NSW Year 2* Units 5 and 6 and *Nelson Maths Building Mental Strategies Skill Book Year 3* Units 12–15, pp. 26–32.

If the student has achieved the recommended skills and these skills are firmly established, consider:

1. Having the student complete *Nelson Maths Building Mental Strategies Skill Book Year 4*, pp. 24–29, to reinforce mental strategies.
2. Moving forward to *Nelson Maths: Australian Curriculum NSW Year 4*, Unit 13.
3. Extending the student by exploring mental addition strategies using fractions and decimals.

Addition

Number and Algebra
Addition and subtraction MA2-5NA uses mental and written strategies for addition and subtraction involving two-, three-, four- and five-digit numbers

add, addition, estimation, MAB, model, ones, ones/tens columns, regroup, rounding up/down, strategy, tens, total

LESSON PLAN 1

TUNING IN

WHAT'S THE TOTAL?

You will need: BLM 15 'Number Cards 1–20'

Give each student a card from BLM 15 'Number Cards 1–20'. Students walk around the room and when you say 'addition', they form a group of three and then add the numbers on their cards. Continue the game, with students forming a different group each time they hear 'addition'. Ask, 'What strategies did you use to work out the total?'

WHOLE-CLASS INTRODUCTION

MODELLING ADDITION WITH MATERIALS

You will need: NTO 3.7 'Modelling with MAB'

Present the equation: 25 + 31. Have students discuss what they need to do to solve this. Ensure they understand that the two amounts need to be combined. Have a student model the numbers 25 and 31 using NTO 3.7 'Modelling with MAB'. Discuss how the ones need to be added and the tens need to be added. Have a student demonstrate how they did this to calculate the total of 56. Present the equation: 46 + 27. Have a student model how they combine the ones and tens using MAB. Ask, 'What should we do now that there are 13 ones and 6 tens?' Discuss and model how the 13 ones can be regrouped as 1 ten and 3 ones. Have students calculate the answer by combining the 7 tens and 3 ones to make 73. Provide further equations.

INDEPENDENT TASKS

Note: Choose from Tasks 1, 2 or 3.

You will need: BLM 8 'Place-Value Mat', 10-sided dice, MAB, BLM 13 'Addition Cards', NTO 3.7 'Modelling with MAB', Student Book p. 24 'Addition in the Garden'

TASK 1: REGROUP TO 50

Give pairs of students BLM 8 'Place-Value Mat', a 10-sided dice and MAB. Students take turns rolling the dice, collecting that number of MAB and placing them on the place-value mat. For example, if a student rolls a 6, they collect 6 ones and place them in the ones column. If the number of ones on the mat becomes greater than 9, the student needs to regroup them as a ten and ones. They move the ten to the tens column and keep the ones in the ones column. The student who regroups to make the number 50 wins the game. Students could also record the addition sums as they go. For example, if there was 24 on the mat and they roll a 3, they record: 24 + 3 = 27.

TASK 2: INTERACTIVE TASK

Give students cards from BLM 13 'Addition Cards' and have them use NTO 3.7 'Modelling with MAB' to model the numbers in the equations. Support students as they add the tens and ones, and regroup if necessary.

TASK 3: STUDENT BOOK p. 24 *'Addition in the Garden'*

TEACHING GROUP

You will need: MAB, NTO 3.7 'Modelling with MAB'

ADDITION WITH NO REGROUPING

- For students who require support, revise addition equations that do not involve regrouping, such as 64 + 23. Have them model the two numbers using NTO 3.7 'Modelling with MAB' or MAB. Encourage students to add the ones together and then add the tens together. Have them solve the equations: 46 + 32, 71 +16, 53 + 41, 64 + 24 and 42 +17. Before attempting addition with renaming, students could revise regrouping. Have students make 16 with ones and support them in regrouping it as 1 ten and 6 ones. Guide students to recognising that the number is still the same amount. Select other numbers for students to regroup.

USING PARTITIONING

- For students who require a challenge, present them with the equation: 46 + 23. Ask, 'What strategies could we use to solve this problem?' Discuss ideas. Use NTO 3.7 'Modelling with MAB' to model 46 and 23. Model how the addition problem could be extended as 40 + 6 + 20 + 3. Group the tens together and the ones together and discuss how it can now be written as 60 + 9. Ask, 'What would be the answer?' Present the equation: 36 + 28. Have a student model it and then record it as 30 + 6 + 20 + 8. Have them add the tens together, then the ones together and then model regrouping the 14 ones as 1 ten and 4 ones. Have the student then record the problem as 60 + 4 and the answer. Provide other problems for students to solve using this strategy.

REFLECTION

Select from the following to suit your class and their learning outcomes:

- Ask, 'Why do you need to regroup when solving addition problems?' Have students identify areas where they regrouped numbers when solving the addition problems and ask them to explain what they did.
- Present the equation: 46 + 39. Ask, 'Would you need to regroup when solving this addition problem? How do you know?' Encourage students to explain how MAB could help them solve this equation.

LESSON PLAN 2

TUNING IN

THE ANSWER IS 20

Write 16 + 4 = 20 on a sheet of paper but do not show it. Write 20 on the board and say that you have written an equation that equals 20 on the paper. Ask, 'What addition problem have I written?' Record suggestions and show students the equation when someone thinks of 16 + 4. Discuss the strategies the students used.

WHOLE-CLASS INTRODUCTION

VERTICAL RECORDING

Present the equation: 46 + 23 (horizontally). Explain that this equation can be recorded in another way. Record the equation vertically and discuss how the tens and ones in each number are lined up vertically. Emphasise how the ones need to be added together first and then the tens. Model how the answer is recorded. Then present the equation: 37 + 28 (horizontally). Ask, 'How would we record this problem vertically?' Have students add the ones and model how the 15 ones are regrouped and recorded as 1 ten and 5 ones. Then add the tens together and record the answer. Have students explain how they would record and solve 46 + 21 and 57 + 25.

INDEPENDENT TASKS

Note: Choose from Tasks 1, 2 or 3.

You will need: BLM 18 'Check This', BLM 13 'Addition Cards', small whiteboards, MAB, NTO 3.11 'Calculator', Student Book p. 25 'Which Number?'

TASK 1: CHECK THIS

Students use their knowledge of how addition problems are recorded and solved vertically to check the equations on BLM 18 'Check This'. Have them identify correct answers and where the errors have been made.

TASK 2: INTERACTIVE TASK 

Give students a set of cards made from BLM 13 'Addition Cards'. Have them record the problems vertically on small whiteboards and solve them. They can use MAB to help with regrouping if necessary. Students can then check their answers using NTO 3.11 'Calculator'. If they made an error when solving the equation, encourage them to revise their work and check how the ones and tens were added, and if they regrouped correctly.

TASK 3: STUDENT BOOK p. 25 ***'Which Number?'***

TEACHING GROUP

You will need: MAB, BLM 13 'Addition Cards', BLM 5 'Digit Cards', BLM 19 'Addition Boxes'

CARD FLIP SUMS

- For students who require support, present the equations: 35 + 13, 42 + 39. Record the problems vertically and solve them. Use MAB to model if necessary. As a group, make a poster recording the steps taken in solving both problems. Emphasise how one problem required regrouping and the other one did not. Ask, 'Why did we need to regroup when solving 42 + 39? What was different about the two problems?'

Present students with cards made from BLM 13 'Addition Cards' and have them record and solve the equations vertically, then sort them into two categories: equations with no regrouping and equations with regrouping. Encourage them to refer to the poster when solving and sorting the problems.

ADDITION PUZZLE GAME

- For students requiring a challenge, have them play an addition puzzle game with a partner. Provide each pair with a set of cards made from BLM 5 'Digit Cards'. Have students share out the cards and then take turns at placing a card in an empty box on BLM 19 'Addition Boxes'. If a student is unable to place a card because it won't 'fit' in the problem, then they miss their turn. The student who puts the last card down to complete the problem wins that game. Students then play the game again, but each time they must have a different equation at the end. Encourage students to make addition equations that involve regrouping.

REFLECTION

Select from the following to suit your class and their learning outcomes:

- Have students explain how they would record and solve 63 + 25 vertically.
- Encourage students to reflect on the addition puzzle game in the Teaching Group. Ask, 'What strategies did you use to help you win? What made you think of where to place different cards?'

LESSON PLAN 3

TUNING IN

UP OR DOWN

You will need: NTO 3.10 'Card Flip'

As a group, revise the rules for rounding numbers to the nearest ten. Then select 0–99 cards from NTO 3.10 'Card Flip' and present numbers to students. Explain that if the number needs to be rounded up, they put their thumbs up and if it needs to be rounded down, they put their thumbs down.

WHOLE-CLASS INTRODUCTION

PARTNER ROUNDING

You will need: BLM 1 '2-Digit Number Cards'

Have students discuss how rounding numbers can help with solving or checking answers to addition problems. Present the equation: 57 + 31. Ask, 'How could rounding numbers help us solve this problem?' Explore steps, including rounding each of the numbers up or down to the nearest ten and then adding those numbers together. Give each student a card from BLM 1 '2-Digit Number Cards'. Have them move around the room; when you say 'stop' they are to find a partner, round their numbers to the nearest ten and add those numbers. Explain how this gives an estimate of the answer. Have students swap cards and repeat the activity.

INDEPENDENT TASKS

Note: Choose from Tasks 1, 2 or 3.

You will need: BLM 1 '2-Digit Number Cards', calculators, small whiteboards, LO: *L94 'The part-adder: generate hard sums'*, Student Book p. 26 'Estimating Problems'

TASK 1: CALCULATOR CHALLENGE

Give pairs of students a set of cards made from BLM 1 '2-Digit Number Cards' and a calculator. The cards are placed face down in a pile between the students. One student turns over two cards and estimates the total by rounding the numbers up or down and then adding those numbers together. They can record their equation and answer on a small whiteboard. The other student works out the answer on the calculator. If the student's estimation is within ten numbers of the correct answer, they win one point. Students then swap roles.

TASK 2: INTERACTIVE TASK 

Have students use LO: *L94 'The part-adder: generate hard sums'* to solve addition equations. Before entering their final answer, have students estimate the total by rounding the numbers in the equation to the nearest ten and adding them. If their answer is not similar to their estimation, students should revise the problem.

TASK 3: STUDENT BOOK p. 26 ***'Estimating Problems'***

TEACHING GROUP

You will need: BLM 13 'Addition Cards', MAB, calculators, decks of playing cards (picture cards and jokers removed)

MODELLING WITH MATERIALS

- For students who require support, have them select a card from BLM 13 'Addition Cards'. Guide students as they round the numbers up or down to the nearest ten and have them use MAB to model these numbers. For example, for 37 + 23, students would collect 4 tens and 2 tens. Students can estimate the answer by adding the tens together, then use the calculator to check the accuracy of their estimation.

ESTIMATE THIS

- For students who require a challenge, have them work with larger numbers. Students turn over five cards from a deck of playing cards and form a 3-digit number and a 2-digit number. For example, from 4, 5, 8, 2 and 3, a student could make 458 and 23. Have students estimate the answer by rounding the numbers and then adding them. Students can then work out the real answer on a calculator. Ask, 'Was your estimate accurate? If your estimation and the answer on the calculator aren't similar, what should you do?'

REFLECTION

Select from the following to suit your class and their learning outcomes:

- Have students share their work from the Independent Tasks or the Teaching Group and discuss how they used the estimation strategy. Ask, 'How can estimation help you solve addition problems? How can rounding numbers help you work out if your answers are accurate?'
- Discuss what students would do if their answer and estimation were not similar.

Home Tasks

Select from the possible Home Tasks:

- Have students look at catalogues (either hard copy or online) and find the total cost of collections of items by adding the prices together. Encourage students to check their accuracy by rounding each number and then adding to find an estimate.
- Have students find 2-digit numbers around the home, e.g. in books. Have them record and solve vertical addition problems using these numbers.

Assessment

- Have students complete **Student Assessment p. 27**.
- Review with students **Assessment Task Card 3.6**.

During the three lessons:

- Identify who can solve addition problems with regrouping and also accurately estimate the answers.
- Make note of students who completed the scaffolding tasks or the more challenging activities of the Teaching Groups.
- Review Student Book pages and make notes of areas of difficulty.

Recommendations for Future Learning

Specific to Student Assessment p. 27; if the student is experiencing some difficulty:

Q 1 Have the student solve the 2-digit with 2-digit addition equations using MAB to model the numbers. Have the student also use MAB to model the regrouping of 10 ones for a ten.

Q 2–3 Discuss with the student the specific steps involved in solving addition equations, such as adding ones, adding tens and regrouping. Encourage the student to think aloud as they solve the problems to ensure they are following all the steps.

Q 4 Have the student revisit how they estimate the answer to addition problems. Have the student practise rounding 2-digit numbers to the nearest ten and then encourage them to add these numbers together.

If the student has not achieved the recommended skills for this unit:

1. See **Assessment Task Card 3.6** for specific recommendations.
2. Have the student add 2-digit numbers without regrouping before moving to addition with regrouping.
3. Review *Nelson Maths: Australian Curriculum NSW Year 2* Units 5 and 6.
4. Review *Nelson Maths Building Mental Strategies Skill Book Year 3*, pp. 40–42, to reinforce using estimating.

If the student has achieved the recommended skills and these skills are firmly established, consider:

1. Having the student complete *Nelson Maths Building Mental Strategies Skill Book Year 4*, pp. 26–28, to reinforce mental strategies for addition.
2. Moving forward to *Nelson Maths Australian Curriculum NSW Year 4* Unit 13.
3. Extending the student in any of the listed activities by using larger numbers or numbers involving decimals.

Unit 7 Position

Measurement and Geometry
Position MA2-17MG uses simple maps and grids to represent position and follow routes, including using compass directions

backwards, bird's-eye view, coordinates, directions, down, forwards, go, grid, left, map, path, right, route, stop, up

LESSON PLAN 1

TUNING IN

POSITION WORDS

Have students brainstorm words they use when they give or hear directions, such as forwards, backwards, go, stop, left, right, up and down. Record the words on the board and discuss their meanings. Tell students to stand up and follow your directions. Say words from the board to move students around the room, e.g. 'Walk forwards. Turn right. Walk backwards. Stop. Sit down.'

WHOLE-CLASS INTRODUCTION

WRITING DIRECTIONS

Tell students to write directions that explain how to get somewhere in the school, such as the library. Ask, 'What type of words should we use in our directions?' Have students discuss the route to the library and record their directions on a sheet of paper, then follow the directions to see if they get to the library. Discuss what happened.

INDEPENDENT TASKS

Note: Choose from Tasks 1, 2 or 3.

You will need: BLM 20 'Directions', traffic cones, LO: *L1074 'Direct a robot: which way?'*, Student Book p. 28 'Writing Directions'

TASK 1: FORWARDS, BACKWARDS, LEFT, RIGHT, STOP

Have students work with a partner using cards made from BLM 20 'Directions' and a traffic cone. Have one student place the cone somewhere in the room, then use the cards to direct their partner from the door to the cone. They are not allowed to talk to their partner. Have students swap roles.

TASK 2: INTERACTIVE TASK

Have students use LO: *L1074 'Direct a robot: which way?'* to give directions to create specific paths.

TASK 3: STUDENT BOOK p. 28 *'Writing Directions'*

TEACHING GROUP

You will need: objects that can be hidden in the classroom (e.g. toy, block, ball), BLM 10 'Grid Paper'

FIND IT

- For students who require support, have them work with a partner. Have one student hide an object in the room, without their partner seeing. They then give their partner directions to find the hidden object. If their partner successfully finds the object, then the student who gave the directions receives one point. Have students swap roles.

DESCRIBE THE PATH

- For students who require a challenge, give them each BLM 10 'Grid Paper' and have them work in pairs. Without their partner seeing, one student selects a starting point and a finishing point on their grid and draws a path between them. The student must then describe the path so their partner can record it on their grid paper. Have students compare their paths to see if they are the same. Discuss what they notice. Have students swap roles.

REFLECTION

Select from the following to suit your class and their learning outcomes:

- Have students reflect on the directions they gave during the Independent Tasks and Teaching Group activities. Ask, 'What was difficult? What things do you need to think about when you give directions?'
- Give students a set of directions that lead to somewhere in the school, but don't tell them the destination. Have students predict where the directions would lead them. Ask, 'Can you visualise the path you would be taking?'

LESSON PLAN 

TUNING IN

LOOK AT THE MAP

You will need: a collection of maps – either hard copy or electronic (e.g. map of a zoo, map of a theme park)

Present a collection of maps and ask, 'What do we use maps for?' Discuss students' responses. Talk about how maps can help us locate places and also help us work out how to get from one place to another. Discuss how a map is a bird's-eye view of a place. Ask students to identify different positions on the maps. As a group, use the maps to explain how you would get from one position to another.

WHOLE-CLASS INTRODUCTION

DRAWING A BIRD'S-EYE VIEW

You will need: a collection of different objects (e.g. cup, pencil, book, football), digital camera

Place a collection of objects on the floor and have students stand up and look down at them. Ask, 'What shape did the objects appear when you saw them from above?' Tell students that this is how items or places are drawn on a map. Ask, 'If we were going to draw a map of the items we have here on the floor, what shape would we draw them?' Discuss students' ideas. Take a photo of the objects from above and have students discuss what they see, then draw a map of the items on the board.

INDEPENDENT TASKS

Note: Choose from Tasks 1, 2 or 3.

You will need: *Word*, Student Book p. 29 'Map of My Bedroom'

TASK 1: MAP OF THE CLASSROOM

Have students draw a map of the classroom. Remind them to consider the shape, size and positioning of objects in the room.

TASK 2: INTERACTIVE TASK 

Tell students that they are going to design and draw a map of a new city. Have them begin by making a list of the things they would like to include in their city, e.g. buildings, parks, train stations, rivers, restaurants. Then have students use the drawing tool in *Word* to draw a map of their city. Students can select or draw appropriate shapes to make a bird's-eye view of the positions.

TASK 3: STUDENT BOOK p. 29 *'Map of My Bedroom'*

TEACHING GROUP

You will need: building blocks, BLM 10 'Grid Paper'

BUILD A CITY

- For students who require support, give them a set of building blocks and have them design a city. Have students describe the positions in their city and draw a map of their city, including all the parts that they built. Encourage students to refer back to their construction and look at it from above. Ask, 'What would the buildings in your city need to look like on the map? Will they look exactly like the ones you built, or different?'

MAP ON GRID PAPER

- For students who require a challenge, give them BLM 10 'Grid Paper'. Explain that they need to draw a map of their house. Ask, 'What things will you need to consider when you draw a map on grid paper? How might the grid paper help you to draw a more accurate map?' Have students label specific areas on their map.

REFLECTION

Select from the following to suit your class and their learning outcomes:

- Have students explain what they need to think about when drawing maps. Ask, 'What makes drawing maps difficult?'
- Have students reflect on the Independent Tasks and the Teaching Group activities. Ask, 'Why do you think maps are drawn from a bird's-eye view?'

TUNING IN

ON A GRID

You will need: masking tape or chalk

Make a large 9 x 9 grid on the floor using masking tape or chalk. Record the letters A to I along the bottom and the numbers 1 to 9 up the side. Model how coordinates can be used to tell where things are on the grid. Have a student stand in a square on the grid and others work out the coordinates of that square. Repeat a few times with different squares. Then give students sets of coordinates, such as C6, and have them find and stand in the appropriate squares. Ask, 'How would coordinates help us use maps?'

WHOLE-CLASS INTRODUCTION

MAP OF THE SCHOOL

Make a list of places at school, such as the buildings, oval and playground. Ask, 'If we were going to draw a map of the school, how would we include each of these places?' Discuss students' ideas and strategies for drawing a map. As a group, draw a map of the school, emphasising the shape and size of each position. Add grid lines and coordinates to the map. Ask, 'How will coordinates help us identify places on our map?' Have students identify the coordinates for specific positions and mark out routes between different positions on the map.

INDEPENDENT TASKS

Note: Choose from Tasks 1, 2 or 3.

You will need: BLM 21 'Shopping Centre Map', LO: *L753 'Direct a robot: collector'*, Student Book p. 30 'Using a Map'

TASK 1: GOING SHOPPING

Give students BLM 21 'Shopping Centre Map'. Explain that they are going shopping and have been given the following shopping list: a loaf of bread, a pair of shoes, a bag of apples, a T-shirt and a teddy bear. Ask students to record the path they would take on the map. Have them write the directions they would give to someone else if they were going to the shopping centre with the same list. Students then make their own shopping lists, draw the path on the map and record the directions.

TASK 2: INTERACTIVE TASK

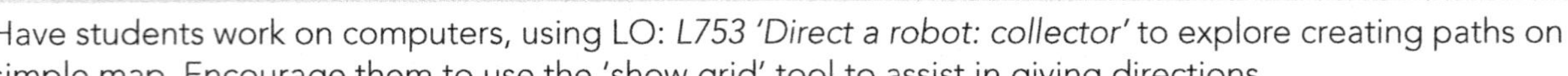

Have students work on computers, using LO: *L753 'Direct a robot: collector'* to explore creating paths on a simple map. Encourage them to use the 'show grid' tool to assist in giving directions.

TASK 3: STUDENT BOOK p. 30 *'Using a Map'*

TEACHING GROUP

You will need: map of the local area with coordinates (e.g. from a local street directory or using an interactive map tool), BLM 10 'Grid Paper'

WHAT CAN YOU FIND?

- For students who require support, have them look at a map of the local area. Ask, 'What could we use this map for?' Discuss students' responses. Identify the coordinates on the map. Ask, 'What could we use these coordinates for?' Have students identify and name the coordinates of local landmarks on the map, such as the school, shopping centres and parks. As a group, mark pathways between different locations on the map. Have students try to find their house on the map and write directions for the route they take to school.

BATTLESHIPS

- For students who require a challenge, give them BLM 10 'Grid Paper' and ask them to write letters along the bottom and numbers up the side of their grid to make coordinates. Students work in pairs and each draw five three-square 'battleships' on their grid without their partner seeing. They take turns to try to find their partner's battleships by naming coordinates. For example, if a student says C8, their partner checks their grid to see if they have a battleship at that position. If there isn't, they say 'miss' and if there is, they say 'hit'. Have students mark their grids accordingly. The first student to 'hit' all their partner's battleships is the winner.

REFLECTION

Select from the following to suit your class and their learning outcomes:

- Have students reflect on the Independent Tasks and Teaching Group activities. Have them explain how coordinates on maps can help them find positions and work out routes from one position to another.
- Have students share their experiences from the battleships game in the Teaching Group. Ask, 'How did using coordinates help you locate your partner's battleships?' Discuss why grids and coordinates are important features of a map.

Home Tasks

Select from the possible Home Tasks:

- Have students follow simple directions from parents or carers, e.g. 'Walk forwards, turn right, look down and get the container off the shelf.' Encourage them to give directions to others.
- Encourage students to read and interpret simple maps at different places. For example, have them locate where shops are on a shopping centre map.
- Encourage students to draw maps of familiar places, such as the kitchen or backyard.

Assessment

- Have students complete **Student Assessment p. 31**.
- Review with students **Assessment Task Card 3.7**.

During the three lessons:

- Observe which students are able to give and follow simple instructions. Identify students who are able to create, interpret and use simple grid maps.
- Make note of students who completed the scaffolding tasks or the more challenging Teaching Group activities.
- Review Student Book pages and make notes of areas of difficulty.

Recommendations for Future Learning

Specific to Student Assessment p. 31; if the student is experiencing some difficulty:

Q 1 Revise following and giving directions. Discuss the words used for giving directions and ensure the student understands their meanings. Give simple directions for the student to follow around the classroom. Then encourage the student to give simple instructions to others.

Q 2 Review strategies for drawing maps. Discuss how objects on maps are drawn from a bird's-eye view. Provide opportunities for the student to draw simple maps and focus on the size, shape and position of different positions.

Q 3 Review simple grid maps and coordinates. Have the student explore how to locate specific positions on the map. Have the student use maps to create and follow pathways.

If the student has not achieved the recommended skills for this unit:

1. See **Assessment Task Card 3.7** for specific recommendations.
2. Have the student draw simple maps of familiar areas.
3. Provide the student with simple grid maps and support them in creating and interpreting positions and pathways.
4. Review *Nelson Maths: Australian Curriculum NSW Year 2* Unit 10.

If the student has achieved the recommended skills and these skills are firmly established, consider:

1. Moving forward to *Nelson Maths: Australian Curriculum NSW Year 4* Unit 12.
2. Providing opportunities for the student to explore compass points and how they relate to maps.

Unit 8 Place Value

Number and Algebra
Whole numbers MA2-4NA applies place value to order, read and represent numbers of up to five digits

expand, hundreds, MAB, ones, partition, place value, regroup, rename, spike abacus, tens, thousands

LESSON PLAN

TUNING IN

MAKING $100

You will need: BLM 22 'Banknotes'

Display the $5, $10, $20 and $50 notes from BLM 22 'Banknotes' and pose the following problem: 'I went to the bank yesterday to withdraw $100, but they didn't have any $100 notes left. What notes might they have given me instead? Can you think of more than one solution?' Observe how students solve this problem and the variety of responses they provide.

WHOLE-CLASS INTRODUCTION

OH, NO! NO HUNDREDS!

You will need: MAB class set with hundreds taken out, BLM 23 '4-Digit Number Expanders'

Write a 4-digit number on the board, e.g. 8257. Select a student to model it with MAB. The student should look for the hundreds and find none. Ask, 'Can we model this number without hundreds? What could we use instead?' Model the number using 10 tens for each hundred. Total the MAB used, one at a time, writing amounts on the board (e.g. '8 thousands, 25 hundreds and 7 ones), and ask, 'Does this model represent our number?' Present a 4-digit number expander to the class and explain that these can be used to expand and rename numbers. Write the numbers in the spaces provided and show the class. Then fold away the hundreds value, to show only thousands, tens and ones. This should represent the class model. Ask, 'What if we had no tens either?' Discuss and fold the tens away, showing only thousands and ones (e.g. 8 thousands, 257 ones). Finally, rename the number only using ones.

INDEPENDENT TASKS

Note: Choose from Tasks 1, 2 or 3.

You will need: BLM 23 '4-Digit Number Expanders', BLM 6 '4-Digit Number Cards', MAB class set, NTO 3.7 'Modelling with MAB', Student Book p. 32 'Using Number Expanders'

TASK 1: USING NUMBER EXPANDERS

Give pairs of students BLM 23 '4-Digit Number Expanders' and five numbers from BLM 6 '4-Digit Number Cards'. Students use the number expander to expand each number, then rename each number in four different ways using different combinations of no thousands, hundreds or tens.

TASK 2: INTERACTIVE TASK

Give students three numbers from BLM 6 '4-Digit Number Cards'. Ask students to use NTO 3.7 'Modelling with MAB' to expand each number, then rename it using no tens, then no hundreds and finally, no thousands. Students may use a number expander to help. Have students record each renamed amount as they create it.

TASK 3: STUDENT BOOK p. 32 *'Using Number Expanders'*

TEACHING GROUP

You will need: MAB, BLM 23 '4-Digit Number Expanders'

MODELLING AND REMODELLING SMALLER NUMBERS

- For students who require support, use MAB to model a 3-digit number. Ask, 'How could we make this number if we had no hundreds?' Emphasise that 10 tens equal 1 hundred, so we would need to add 10 tens for every hundred needed. Repeat as necessary, modelling with no tens or no hundreds. Demonstrate on the number expander. Move on to thousands when ready.

CHALLENGING RULES

- For students who require a challenge, ask them to generate five different 4-digit numbers and model them one at a time with MAB. Say, 'You are required to model your first number using no thousands and only 2 tens.' For the remaining four numbers, students can make up their own rules for renaming, then swap with a partner and complete each other's challenges.

REFLECTION

Select from the following to suit your class and their learning outcomes:

- Write the number 1 273 on the board, and ask, 'How would you rename this number with no hundreds or thousands?'
- Ask, 'What would happen if you had no ones? Could you make the number? What might you do?'

LESSON PLAN 

TUNING IN

MAKING NUMBERS

You will need: NTO 3.12 'Stopwatch' or a stopwatch

Display NTO 3.12 'Stopwatch' on the IWB and set to count down from one minute (or use an actual stopwatch). Ask, 'How many numbers can you make with a 5 in the hundreds place in one minute?' Note whether students use 3-digit numbers or if they move into the thousands.

WHOLE-CLASS INTRODUCTION

PARTITION

You will need: NTO 3.2 'Extended Notation: Number Cards'

Display NTO 3.2 'Extended Notation: Number Cards' on the IWB with the 'show MAB' option selected. Select a 3-digit number from Tuning In, e.g. 531, and write it on the screen. Ask a student to name this number and model it using MAB. Explore how to partition the number. Ask, 'How many hundreds are there? What are they worth?' Select the 500 number card and place it below the MAB hundreds. Then ask, 'How many tens? What are they worth?' Add the + symbol and the 30 card to show addition of parts. Repeat for the ones, e.g. 500 + 30 + 1. Extend understanding by repeating the process using a variety of 4-digit numbers from Tuning In.

INDEPENDENT TASKS

Note: Choose from Tasks 1, 2 or 3.

You will need: car sales section of newspaper, BLM 7 'MAB', access to the internet, NTO 3.2 'Extended Notation: Number Cards', *PowerPoint*, Student Book p. 33 'Cars for Sale!'

TASK 1: CARS FOR SALE

Give students one page of the car sales section of the newspaper and ask them to find five cars with a price of less than $10 000. They cut out each car and stick it in their workbooks. Beside each, they partition the value of each car and model it with MAB using BLM 7 'MAB'.

TASK 2: INTERACTIVE TASK 

Have students look at an online car sales website. In the search criteria, limit the price range to a maximum of $10 000. Ask students to find ten prices that have a 9 in the thousands column and to use NTO 3.2 'Extended Notation: Number Cards' to model the partition of each number. If time permits, students could create a *PowerPoint* presentation showing the cars in order of value and their partitioned amounts.

TASK 3: STUDENT BOOK p. 33 *'Cars for Sale!'*

TEACHING GROUP

You will need: MAB, BLM 8 'Place-Value Mat', car sales section of newspaper

EXTENDED FORMAT

- For students who require support, ask them to show you the numbers they generated in Tuning In, and select a 3-digit number from these. Give each student BLM 8 'Place-Value Mat' and ask them to model their number on the mat with MAB. Below the MAB, have them partition the number by writing the value of each part. Draw out links between the model, its numeric representation and the partitioning.

CLASSY WHEELS!

- For students who require a challenge, give them a page of the car sales section of the newspaper and ask them to find ten cars that are priced between $15 000 and $20 000. Ask students to order these and write each amount in partitioned format. Students then create a scaled number line to fit their numbers.

REFLECTION

Select from the following to suit your class and their learning outcomes:

- Repeat the task from Tuning In with a different numeral in the hundreds place. Assess whether students are able to generate more numbers.
- Ask, 'How does partitioning a number help us to understand the value of the number?'

LESSON PLAN 3

TUNING IN

MY PHONE NUMBER

You will need: BLM 24 'Phone Number Hunt'

Give students BLM 24 'Phone Number Hunt'. Have them write the last four digits of their phone number at the top of the page, then move around the room to find other people who have the right numbers to fill all the spaces on their sheet.

WHOLE-CLASS INTRODUCTION

MODELLING AND REGROUPING

You will need: NTO 3.7 'Modelling with MAB', MAB, spike abacus

Display NTO 3.7 'Modelling with MAB' on the IWB with the 'show thousands' option selected. Select one student to write the last four digits of their phone number on the board. Ask, 'How could we model this?' Discuss responses and model using partitioning, MAB and the spike abacus. Ask, 'What would a number 20 more than this look like?' Model with each method. Repeat by adding different amounts of tens, hundreds and thousands, e.g. +/–20, +/–300, +/–4000. Provide examples that require regrouping, e.g. adding 4 ones when there are already 8 will make 1 extra ten and 2 ones. Provide an example that regroups from the thousands into the tens of thousands. Ask, 'If we added another column next to the thousands, what would its value be? How would we model it?' Have students discuss ways to model using partitioning, MAB and the spike abacus.

INDEPENDENT TASKS

Note: Choose from Tasks 1, 2 or 3.

You will need: BLM 8 'Place-Value Mat', MAB, spike abacus, dice (1 per pair), *PowerPoint*, Student Book p. 34 'Number Hunt'

TASK 1: FIRST TO 10000

Give pairs of students a copy of BLM 8 'Place-Value Mat' each, and have them share a spike abacus or MAB, and a dice. Each student starts with their own 4-digit phone number, written on the mat and modelled with the abacus or MAB. Students take turns to roll the dice and from the digit generated, select to which place-value column they will add or subtract that amount, e.g. if a 6 is rolled it could be worth 6, 60, 600 or 6000. The aim is to be the first to reach exactly 10000. Note: students may use calculators.

TASK 2: INTERACTIVE TASK 

Have students create a *PowerPoint* presentation using five of the 4-digit phone numbers from Tuning In. Each number is to be placed in the centre of a new slide. In each of the four corners surrounding the number, students are to: write the number in words, partition the number, write the number that is 5000 more than their number and write a place-value fact about their number.

TASK 3: STUDENT BOOK p. 34 *'Number Hunt'*

TEACHING GROUP

You will need: MAB or spike abacus

ADDING 10, 20 AND 30

- For students who require support, play a counting game where you begin at 100 and count by 10s. Once understood, move on to 20s and then 30s. As the numbers near the next hundred, model the number with MAB or a spike abacus and show the amount that is being added and how this regroups to the next column. Once consolidated, move on to the hundreds and then the thousands.

ADDING 30, 300, 3000

- For students who require a challenge, play the same counting game as above but start with a 4-digit number and add 30, 300 or 3000. These numbers will not be modelled, but students may require pen and paper to work them out. Provide a time limit (say, 10 seconds) to answer. Students whose answers are incorrect are out. The winner is the last student remaining. As an extra challenge, use subtraction.

REFLECTION

Select from the following to suit your class and their learning outcomes:

- Invite students who completed the *PowerPoint* activity in Independent Tasks, Task 2, to present them to the class.
- Have students stand in a circle. Start with any 3-digit number and create a number pattern by going around the circle and having each student add 20 to the number.

Home Tasks

Select from the possible Home Tasks:

- Have students show their parents or carers how to rename numbers when you have no tens, hundreds or thousands. Have them create renaming challenges for each other and work together to solve them. Note: some students may wish to take a copy of BLM 23 '4-Digit Number Expanders' home to help them.
- Have students find four more car advertisements that suit their criteria from the Independent Tasks in Lesson Plan 2. Have them cut or print out the advertisements, partition the amounts and bring them to school to add to their work.

Assessment

- Have students complete **Student Assessment p. 35**.
- Review with students **Assessment Task Card 3.8**.

During the three lessons:

- Take photos of students working on the renaming activities in Lesson Plan 1 to add to students' portfolios. Students could be asked to reflect on the activity and their learning.
- Collect digital material such as the *PowerPoint* presentations from Lesson Plans 2 and 3 as work samples for digital portfolios.
- Make note of how students go about solving the Tuning In activities in Lesson Plans 1 and 2 and whether their knowledge develops throughout the lessons.

Recommendations for Future Learning

Specific to Student Assessment p. 35; if the student is experiencing some difficulty:

Q 1 Revise the expanding and renaming activities in Lesson Plan 1, looking particularly at the scaffolding tasks.

Q 2 Revise the place-value chart and modelling with MAB to gain an understanding of the value of digits in a number.

Q 3 Revisit the 'Partition' activity from the Whole-Class Introduction in Lesson Plan 2. Model numbers with MAB and link this to partitioning.

Q 4 Have the student practise adding and subtracting 10, 20 and 30 from any 3-digit number. Model the number so the student has concrete materials to manipulate in order to reach an answer.

If the student has not achieved the recommended skills for this unit:

1. See **Assessment Task Card 3.8** for specific recommendations.
2. Have the student work with 2- and 3-digit numbers before moving on to larger numbers. Scaffold the student with the use of MAB, number expanders and other concrete materials to reinforce the value of numbers.
3. Review *Nelson Maths: Australian Curriculum NSW Year 3* Units 1 and 2.

If the student has achieved the recommended skills and these skills are firmly established, consider:

1. Having the student complete *Building Mental Strategies Skill Book NSW Year 3*, pp. 16–17.
2. Moving forward to *Nelson Maths: Australian Curriculum NSW Year 4* Unit 3.
3. Extending the student in any of the listed activities by using larger numbers.

Unit 9 More About Place Value

Number and Algebra
Whole numbers MA2-4NA applies place value to order, read and represent numbers of up to five digits

addition, hundreds, ones, place value, regrouping, subtraction, tens, thousands

LESSON PLAN 1

TUNING IN

MYSTERY CODE

Write the following sequence on the board:

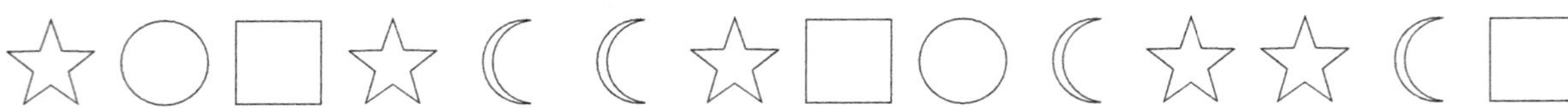

Tell students that this sequence represents a number. Ask, 'What could it be?' Allow time for students to examine and think, then share responses. Tell students that in this sequence, each symbol holds a value, just like the different MAB materials. For example, a star may be worth 10; there are five stars so the number would have a 5 in the tens place. Ask students to work out four possible values for the sequence. Each number should contain the digits 2, 3, 4 and 5.

WHOLE-CLASS INTRODUCTION

A CLASS CODE

You will need: BLM 8 'Place-Value Mat' or a 4-digit mat drawn on the board, BLM 25 'Number Hieroglyphs'

Tell students that the number represented in Tuning In is 2453. Ask, 'Did anyone guess this number?' If so, ask them to explain the value of each symbol (circle: thousands, moon: hundreds, star: tens, square: ones). Write the symbols on a place-value mat so that students can see their value and the number of symbols in each column to show 2453. Emphasise the link between these symbols and modelling materials such as MAB or the spike abacus. Using this class code, have students create a picture sequence for any 4-digit number. Once complete, have students swap with the person next to them to 'crack' their code. Display BLM 25 'Number Hieroglyphs' on the board. Ask, 'Has anyone seen these before? What are they? How are they similar to our class code?' Discuss, drawing out links that all have values for 1, 10, 100 and 1000. Examine the other digits included in the hieroglyphs. Discuss possible reasons for their inclusion.

INDEPENDENT TASKS

Note: Choose from Tasks 1, 2 or 3.

You will need: BLM 25 'Number Hieroglyphs', BLM 6 '4-Digit Number Cards', access to the internet, Student Book p. 36 'Ancient Egyptian Hieroglyphs'

TASK 1: A CODE OF MY OWN

Using the class code and BLM 25 'Number Hieroglyphs' (displayed on the board) as inspiration, have students create their own number code, drawing pictures for the digits 1–10, 100 and 1000 (and 50 and 500 if desired). Have students write their code at the top of a page and draw a box around it. They then create five different equations all written in code, beginning with a 4-digit number and adding or subtracting a chosen amount, e.g. 2471 – 500. Have students swap with a partner and solve each other's equations.

TASK 2: INTERACTIVE TASK 

Have students search the internet for information on Roman numerals. Give each student five 4-digit number cards from BLM 6 '4-Digit Number Cards' to write using Roman numerals. Ask students to subtract 500 from each number and write the new value in Roman numerals.

TASK 3: STUDENT BOOK p. 36 *'Ancient Egyptian Hieroglyphs'*

TEACHING GROUP

You will need: MAB, laminated copies of BLM 8 'Place-Value Mat', whiteboard markers, BLM 6 '4-Digit Number Cards'

THE CODE BASICS

- For students who require support, work as a group to create a new code. For each value, show students the corresponding MAB and ask, 'Instead of MAB, what could we use to represent our [thousands]?' Draw a master copy of the code on one of the place-value charts. Give each student a whiteboard marker and a laminated copy of BLM 8 'Place-Value Mat'. Select one card from BLM 6 '4-Digit Number Cards' and ask students to model the number using their code. Assist as required.

AN ANCIENT EGYPTIAN CHALLENGE

- For students who require a challenge, extend Independent Tasks, Task 1, by having them develop their own code to include values for 10000 and 100000. Have them create six numbers (two 4-digit, two 5-digit and two 6-digit) and write them in code. Students then create four different equations using only the numbers they have created and swap with a partner to solve.

REFLECTION

Select from the following to suit your class and their learning outcomes:

- Select one or two students to write their code on the board and one of their code numbers/equations. Have the class solve it.
- Ask, 'Why do you think we no longer use Roman numerals or hieroglyphs? What are the benefits of our number system?'

LESSON PLAN 2

TUNING IN

SKIP COUNTING

You will need: four large dice

Have students stand in a circle. Roll four large dice in the centre of the circle and select one student to arrange them into a 4-digit number and say the number aloud. Then, moving around the circle, have students skip count by adding 10, 100 or 1000. Then count backwards by subtracting 10 or 100.

WHOLE-CLASS INTRODUCTION

BEAT THE TEACHER

You will need: A3 copy of BLM 26 'Place-Value Ladders', four dice

Select one student to play against the teacher and model this game to the class. Attach an A3 copy of BLM 26 'Place-Value Ladders' to the board. The aim is to keep creating bigger numbers to climb the ladder. Start by rolling four dice. Make the smallest number you can with these digits (you do not need to explain this strategy to students) and write your number on the first rung of your ladder. The student then has their turn, creates a number and writes it on the first rung of their ladder. The game continues, with players taking turns. Watch out for the 'rung stoppers' at the side, which must be added to or subtracted from the number while still maintaining a larger number than the last. If a player is unable to create a larger number with the digits rolled, they are out.

INDEPENDENT TASKS

Note: Choose from Tasks 1, 2 or 3.

You will need: BLM 26 'Place-Value Ladders', dice (four for each pair), *PowerPoint*, Student Book p. 37 'Place-Value Path'

TASK 1: PLACE-VALUE LADDERS

Give pairs of students BLM 26 'Place-Value Ladders' and have them play the game (see the Whole-Class Introduction for instructions). Once students have played one game, have them modify the existing 'rung stoppers' and add two more of their own, then play again. The game could also be modified to start with the largest number on the top and work down the ladder.

TASK 2: INTERACTIVE TASK 

Have students create a *PowerPoint* presentation beginning with a 4-digit number of their choice. They must perform ten different additions or subtractions to their number, each on a new slide, to show the progression. Students should show evidence of regrouping through their equations. Difficulty of equations may vary depending on student ability.

TASK 3: STUDENT BOOK p. 37 ***'Place-Value Path'***

TEACHING GROUP

You will need: BLM 26 'Place-Value Ladders', MAB or spike abacus

3-DIGIT NUMBER LADDERS

- For students who require support, play BLM 26 'Place-Value Ladders' using three dice to create 3-digit numbers. Have students model each number with MAB or a spike abacus to ensure the next number is larger. Discuss strategies, such as beginning by making the smallest possible number and slowly making numbers larger.

DESIGNER GAMES

- For students who require a challenge, have them design a new game based on BLM 26 'Place Value-Ladders'. They may wish to include 5-digit numbers and more challenging equations that involve regrouping. Have them play their games with a partner.

REFLECTION

Select from the following to suit your class and their learning outcomes:

- Repeat the Tuning In activity, this time saying, 'Skip count, adding 5 ones [or 3 tens].'
- Have students from Independent Tasks, Task 2, share their *PowerPoint* presentations with the class, describing the changes made to the number in each slide.

LESSON PLAN 3

TUNING IN

WISHBALL

You will need: LO: *L867 'Wishball: whole numbers'*

Play LO: *L867 'Wishball: whole numbers'* as a class. Have a student operate the spinner to generate random digits, then have students work together to decide which place-value column to add to or subtract from as they work towards the target. They use the wishball to select the final digit to reach the target in one turn.

WHOLE-CLASS INTRODUCTION

DOMINO TARGET

You will need: dominoes or BLM 27 'Dominoes'

Using dominoes or BLM 27 'Dominoes', select two dominoes at random and arrange them side by side so that they can be read as a 4-digit number, e.g. 6124:

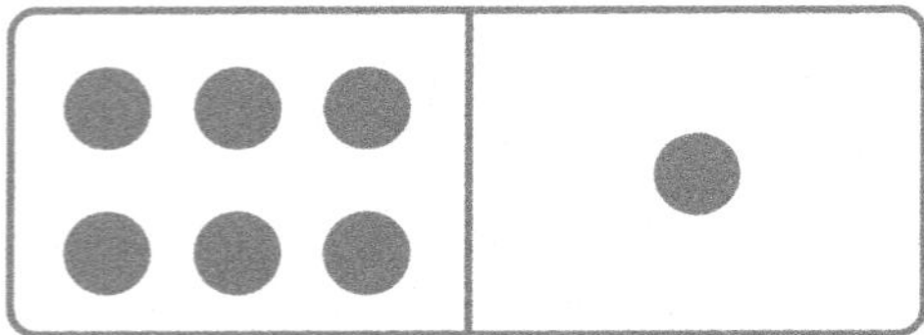

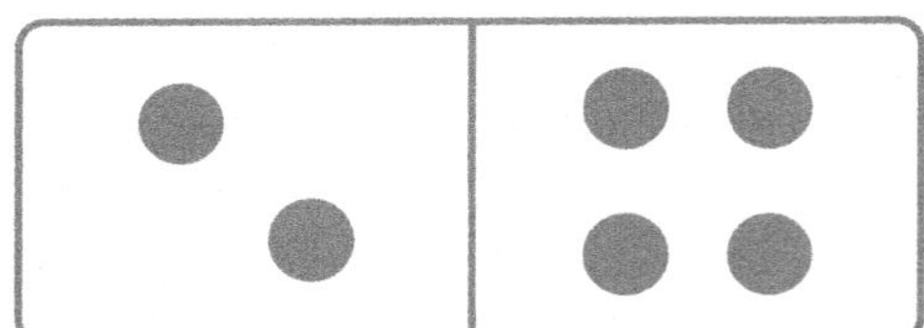

Ask, 'What number is this showing?' Select one student to say the number and write it on the board. Explain that this number is now the target. Invite a student to randomly select two dominoes and arrange them to make a 4-digit number as close to the target as possible. Ask, 'What other numbers could we have made? Is this the closest we can get to the target?' Try another combination of numbers to see if it is closer.

INDEPENDENT TASKS

Note: Choose from Tasks 1, 2 or 3.

You will need: dominoes or BLM 27 'Dominoes', dice, LO: *L871 'Wishball challenge: whole numbers'*, Student Book p. 38 'Reach the Target'

TASK 1: DOMINO TARGET CHALLENGE

Give pairs of students a set of dominoes (or BLM 27 'Dominoes') to play 'Domino Challenge'. Have one student select two dominoes at random to make the target number and write it in words on a sheet of paper. Each student then selects two dominoes and arranges them to make a 4-digit number as close to the target as possible. They then take turns to roll a dice and select the place-value column to which they will add or subtract that amount, e.g. if a 6 is rolled it could be worth 6, 60, 600 or 6000). The winner is the first student to hit the target.

TASK 2: INTERACTIVE TASK 

Have students work independently on computers to play LO: *L871 'Wishball challenge: whole numbers'*.

TASK 3: STUDENT BOOK p. 38 *'Reach the Target'*

TEACHING GROUP

You will need: dominoes or BLM 27 'Dominoes'

REACHING THE TARGET

- For students who require support, simplify the 'Domino Challenge' game by selecting two dominoes and arranging them to create a 4-digit target number. Then have students look at the remaining dominoes and arrange them to make numbers as close to the target as possible. Can they use other dominoes to make the same number? How close can they get?

DOMINO ADDITION AND SUBTRACTION

- For students who require a challenge, play a more challenging game of 'Domino Challenge'. As before, use two dominoes to create a 4-digit target number. Students randomly select four dominoes and arrange them to make two 4-digit numbers, then either add to or subtract from them to get as close to the target as possible. Play the game as a group challenge rather than as a pair.

REFLECTION

Select from the following to suit your class and their learning outcomes:

- Discuss successful strategies used to get close to the target in the domino challenges. Ask, 'How did you select your two 4-digit numbers? What do you know about place value that helped with this activity?'
- Ask students to create individual brainstorm maps showing everything they know about place value.

Home Tasks

Select from the possible Home Tasks:

- Have students create their own place-value game, using dice, dominoes or playing cards. Their game could focus on creating bigger or smaller numbers, ordering numbers or using place value for addition and subtraction. Have them play the game with family members to test effectiveness.
- Have students research and write the numerals from 1–10, 100 and 1 000 in a different language.

Assessment

- Have students complete **Student Assessment p. 39**.
- Review with students **Assessment Task Card 3.9**.

During the three lessons:

- Observe students while they are playing the games in Lesson Plans 2 and 3 and make note of strategies used.
- Collect and take photographs of the place-value games students have created to add to students' digital portfolios. Have students reflect on the effectiveness of the games in their portfolios.
- Review Student Book pages and note areas of difficulty.

Recommendations for Future Learning

Specific to Student Assessment p. 39; if the student is experiencing some difficulty:

Q 1–2 Revisit 'A Class Code' from Lesson Plan 1. Using MAB and a place-value mat, place MAB on the mat and draw the corresponding code symbol in the correct column. Model numbers with both MAB and the code.

Q 3 Have the student model the numbers using MAB, add the required amount and record the new number made. You may also wish to place MAB on a place-value mat to aid understanding.

Q 4 Revise the dominoes activities in Lesson Plan 3. Begin with three 3-digit numbers to simplify the task.

If the student has not achieved the recommended skills for this unit:

1. See **Assessment Task Card 3.9** for specific recommendations.
2. Simplify any of the activities/games from the lessons to 2- and 3-digit numbers, focusing on adding and subtracting 10, 100 and 1 000 before moving on to more difficult equations.
3. Review *Nelson Maths: Australian Curriculum NSW Year 3* Units 1, 2 and 8.

If the student has achieved the recommended skills and these skills are firmly established, consider:

1. Having the student complete *Building Mental Strategies Skill Book Year 3*, pp. 16–17.
2. Moving forward to *Nelson Maths: Australian Curriculum NSW Year 4* Unit 3.
3. Extending the student in any of the listed activities by using larger numbers.

Unit 10 Mass

Measurement and Geometry
Mass MA2-12MG measures, records, compares and estimates the masses of objects using kilograms and grams

equal in mass, gram, heavier, hefting, kilogram, lighter, mass, scales

LESSON PLAN 1

TUNING IN

WHICH IS HEAVIEST?

You will need: classroom items of varying mass, e.g. stapler, pack of pencils, tennis ball, tape dispenser

Select two items and hold them up in front of the class. Have students predict which is heaviest. Discuss. Repeat, comparing three or four pairs of items. Collate responses to make a list of prediction statements on the board, e.g. 'We predict that a stapler is heavier than a tennis ball.'

WHOLE-CLASS INTRODUCTION

USING KILOGRAMS

You will need: beam balance, kilogram mass (e.g. 1 kg weight or 1 L milk carton filled with water)

Ask, 'What is mass? What are we doing when we find the mass of an object?' Discuss to assess prior knowledge. Ask, 'When might we need to find the mass of an object?' List suggestions, e.g. shopping, cooking. Ask, 'What do we use to measure mass?' Discuss kilograms and when we use them, e.g. body weight, grocery items. Ask students to look around the classroom to find examples of things they think would be lighter than, about the same as, or heavier than a kilogram. Ask, 'How could we test these predictions?' Discuss suggestions, e.g. hefting, using a beam balance or scales. Present the beam balance to the class and discuss how it works: the heaviest item pushes the beam down. Invite students to test one of their predictions using the kilogram mass on one side and their classroom item on the other. Repeat as necessary.

INDEPENDENT TASKS

Note: Choose from Tasks 1, 2 or 3.

You will need: beam balance, kilogram mass, *PowerPoint*, access to the internet, Student Book p. 40 'Which Is Heaviest?'

TASK 1: KILOGRAMS – EQUAL, MORE OR LESS

Have students look around the room and list five things they think would weigh about a kilogram, five things that would weigh less and five things that would weigh more. Students then weigh the items using a beam balance with a 1 kg weight on one side to check their estimates.

TASK 2: INTERACTIVE TASK

Have students work in pairs on computers using *PowerPoint* to make a presentation about comparing mass. Have students search the internet to find pictures of items and compare two on each slide, e.g. a cat and a car. Students write statements below the pictures, e.g. 'A cat is lighter than a car. A car is heavier than a cat.'

TASK 3: STUDENT BOOK p. 40 *'Which Is Heaviest?'*

TEACHING GROUP

You will need: classroom items of varying mass, beam balances, uniform units (e.g. MAB, Unifix blocks, dice, counters, felt pens), kilogram mass

INFORMAL COMPARISON

- For students who require support, spend more time discussing the meaning of mass and how we compare items. Present students with two obviously unequal items, e.g. a pencil and a tape dispenser. Ask, 'Which is heavier?' Have them heft the items to feel the difference, then compare them on the beam balance.

Once understood, have students find items that they think will have a similar mass, e.g. a pencil and a highlighter pen, and compare these on the beam balance.

COLLECTIONS EQUAL TO A KILOGRAM

- For students who require a challenge, provide them with a 1 kg weight and a set of uniform units, e.g. Unifix blocks. Have them predict how many Unifix blocks would have a mass of 1 kg, then use the beam balance to find the actual amount. Repeat with another set of uniform units, e.g. MAB, dice, counters or felt pens, to see if predictions become more accurate over time.

REFLECTION

Select from the following to suit your class and their learning outcomes:

- Have students share their experiences with using the beam balance. Ask, 'How did you know whether items were heavier/lighter than a kilogram? Was it difficult to make the beam balance level?'
- Ask, 'Would most items in our classroom weigh more than, less than or equal to one kilogram? Why?'

LESSON PLAN 2

TUNING IN

PENCIL CASE LINE-UP

You will need: pencil cases

Have four students volunteer their pencil cases – ensure they are of different size and mass. Display the pencil cases in front of the class and label them A, B, C and D. Have students predict their order from lightest to heaviest by eye and record their predictions. Discuss. Allow students to heft each pencil case and alter their predictions if required. Discuss and collate predictions to create a 'class order'. Move pencil cases into the predicted order.

WHOLE-CLASS INTRODUCTION

USING GRAMS

You will need: beam balance, class set of gram weights, kilogram mass, pencil cases from Tuning In

Present the beam balance and a set of gram weights. Hold up the weights and ask, 'Do you know what these are?' Hand the weights to students and have them look at the markings. Ask, 'What is written on the weights?' Discuss. Have students think back to the kilogram activities from Lesson Plan 1 and ask, 'How many grams are there in one kilogram?' Use the beam to balance the kilogram mass with a collection of gram weights to show that 1 000 g = 1 kg. Ask, 'How could we measure the mass of our pencil cases? Would they be heavier or lighter than one kilogram?' Discuss. Weigh each pencil case in turn, as accurately as possible, recording the mass of each in grams. Look at the actual mass of each item to decide if the predicted order of pencil cases in Tuning In was correct. If needed, alter the order so the cases are lined up in actual order from lightest to heaviest.

INDEPENDENT TASKS

Note: Choose from Tasks 1, 2 or 3.

You will need: BLM 28 'Mass Cards', supermarket catalogues, LO: *L10570 'HOTmaths: using a beam balance'*, Student Book p. 41 'Weighing in Grams'

TASK 1: SUPERMARKET MASS HUNT!

Give pairs of students a set of cards made from BLM 28 'Mass Cards' and a supermarket catalogue each. Have them set the cards face down on the table. One student turns over a card, and the pair race to see who can find five items in their catalogue that have this mass (or within 50 g of the mass, if necessary). The fastest student receives one point. The winner is the student with the most points after all cards are used.

TASK 2: INTERACTIVE TASK

Have students work independently on computers to play LO: *L10570 'HOTmaths: using a beam balance'* to estimate the mass of everyday objects and check using a beam balance.

TASK 3: STUDENT BOOK p. 41 *'Weighing in Grams'*

TEACHING GROUP

You will need: a selection of supermarket items measured in grams with weights covered, including one item weighing exactly 500 g (e.g. a block of cheese or a tub of margarine); beam balance; supermarket catalogue

500 GRAMS – HEAVIER OR LIGHTER?

- For students who require support, have them hold a supermarket item weighing 500 g to get a feel for this weight. Then, have them hold all other items and put them into three categories according to weight: 'Less than 500 g', 'About 500 g' and 'More than 500 g'. Then weigh each on the beam balance to check. Finally, reveal each weight shown on the item and compare results.

ADDING WEIGHTS

- For students who require a challenge, give them a supermarket catalogue and have them combine items measured in grams to make a total weight of 5 kg. Have them challenge themselves to see how many different combinations they can create to equal 5 kg.

REFLECTION

Select from the following to suit your class and their learning outcomes:

- Have students write a personal reflection on the 'Pencil Case Line-Up' activity in Tuning In. Did their prediction change from using their eye to hefting? How accurate were their predictions? How could their learning help them solve similar problems in the future?
- Play LO: *L10570 'HOTmaths: using a beam balance'* on the IWB as a class.

LESSON PLAN 3

TUNING IN

MILK CARTON MASS

You will need: four milk cartons, sand

Before the class, fill milk cartons with different amounts of sand, label them A, B, C and D and close them so that students cannot see the amount of sand in each. Have students predict the order of mass (lightest to heaviest) by eye and then by hefting. Discuss so that students realise items can look the same yet have different mass.

WHOLE-CLASS INTRODUCTION

USING SCALES

You will need: kitchen scales, sand-filled milk cartons from Tuning In

Present a set of kitchen scales and ask, 'Where have you seen these before? What are they used for? How are they different to the beam balance?' Discuss. Look at the markings on the kitchen scales to emphasise the relationship between grams and kilograms (1 000 g = 1 kg). Place one of the sand-filled milk cartons on the scales, have students watch the pointer move and discuss how to read scales. Select a student to read the weight. Ask, 'What does it weigh? Is this more or less than one kilogram?' Record the exact weight. Repeat for the other sand-filled milk cartons.

INDEPENDENT TASKS

Note: Choose from Tasks 1, 2 or 3.

You will need: kitchen scales, empty 1 L bottles or plastic bowls, sand or rice, BLM 28 'Mass Cards' (excluding top row with kilogram amounts), access to the internet, *PowerPoint*, Student Book p. 42 'Reading and Using Scales'

TASK 1: MAKING WEIGHT

Have students work in pairs with a set of kitchen scales, a 1 L bottle or plastic bowl, sand or rice and a set of BLM 28 'Mass Cards' (with gram amounts only) between them. Have them select a mass card then fill their bottle or bowl with the amount of sand or rice they predict will be needed to create the specified mass. Then, test their prediction on the kitchen scales, reading the mass shown on the scale and deciding whether they need more or less sand or rice. Have them perform alterations as required until the exact mass is reached. Select another mass card and repeat.

TASK 2: INTERACTIVE TASK

Have students work in pairs on computers to research the ideal weight for dog breeds in kilograms (sites such as www.adelaidevet.com.au have this information). Have them select ten breeds and write down their ideal weight (or range), then order them from lightest to heaviest. Have them find a photo of each breed and create a *PowerPoint* presentation of their information.

TASK 3: STUDENT BOOK p. 42 *'Reading and Using Scales'*

TEACHING GROUP

You will need: kilogram weight, beam balance, classroom or supermarket items, kitchen scales, A3 paper

REVISING THE KILOGRAM

- For students who require support, have them work in pairs to find combinations of classroom or supermarket items that equal about 1 kg by placing the kilogram weight on one side of the beam balance and adding and removing smaller items until the beam is level. Have them record as many 'kilogram combinations' as they can. Once the idea of a kilogram is established, they may use kitchen scales to weigh items.

BIGGER BUT LIGHTER, SMALLER BUT HEAVIER

- For students who require a challenge, have them find and weigh classroom items to create five of each of the following statements: a) _____ is bigger than _____, but lighter. b) _____ is smaller than _____, but heavier. Then have them explain how this is possible and present their information on an A3 poster.

REFLECTION

Select from the following to suit your class and their learning outcomes:

- Ask, 'Where else do we use scales? What other kinds of scales have you used? Why do we have different scales?'
- Have students who worked on Independent Tasks, Task 2, share their *PowerPoint* presentations with the class.

Home Tasks

Select from the possible Home Tasks:

- Have students list as many items in their home as they can with 1 kg marked on them, e.g. pantry items, bathroom scales, laundry powder.
- Have students select a recipe and rewrite all of the ingredients, highlighting those that are measured in grams.

Assessment

- Have students complete **Student Assessment p. 43**.
- Review with students **Assessment Task Card 3.10**.

During the three lessons:

- Take photos of students working with the beam balance and kitchen scales to add to their digital portfolios. Students could be invited to write a reflection on the activity and their learning.
- Collect students' *PowerPoint* presentations from Lesson Plans 1 and 3 to show the use of ICT to enhance leaning.
- Make note of any difficulties students have with the Student Book activities.

Recommendations for Future Learning

Specific to Student Assessment p. 43; if the student is experiencing some difficulty:

Q 1 Revise the meaning of 'heaviest'. Revisit activities from Lesson Plan 1 where students compared objects by weight.

Q 2 Revise activities from Lesson Plan 1 where students were introduced to the kilogram. Remind the student of the types of items within the classroom that weighed less than, equal to or more than 1 kg.

Q 3 Revisit the activities from Lesson Plan 2. Have the student think about classroom items they weighed in grams and where they see grams in the supermarket.

Q 4 Remind the student of the meaning of 'lightest' and 'heaviest' and have them compare grams and kilograms. Provide concrete materials if required.

Q 5 Revise activities from Lesson Plan 3 where students read scales. Link to the use of number lines. Provide actual scales for the student to look at if necessary.

If the student has not achieved the recommended skills for this unit:

1. See **Assessment Task Card 3.10** for specific recommendations.
2. Have the student further their understanding of heavier and lighter by comparing two items before moving on to ordering collections of items.
3. Have the student continue to use hands-on measuring devices, such as the beam balance, where they can add weights to emphasise the idea of equal weight.
4. Review *Nelson Maths: Australian Curriculum NSW Year 2* Unit 13.

If the student has achieved the recommended skills and these skills are firmly established, consider:

1. Having the student convert weights (grams to kilograms and vice-versa).
2. Moving forward to *Nelson Maths: Australian Curriculum NSW Year 4* Unit 5.
3. Having the student complete any of the listed activities using larger objects and deciding for themselves which measuring device is most appropriate.

Unit 11 Our Community – Data

Statistics and Probability

Data MA2-18SP selects appropriate methods to collect data, and constructs, compares, interprets and evaluates data displays, including tables, picture graphs and column graphs

axes, axis, categories, collect, column (bar) graph, data, interpret, list, picture graph, represent, table, tally, title

LESSON PLAN 1

TUNING IN

OUR FAVOURITE SONGS

You will need: sticky notes, NTO 3.13 'Data Collection Table'

Have each student write their favourite song on a sticky note and stick it anywhere on the board. Ask, 'What is the most popular song in our class?' Have students attempt to answer. Ask, 'Is it difficult to find which song we like best? Why?' Discuss the need to sort our data. Ask, 'How could we sort it?' Discuss the idea of putting the same songs together. Select students to move the sticky notes into groups. Examine the data and select the five most popular songs. Present NTO 3.13 'Data Collection Table' and explain how tables can be used to collect and sort data. Type the heading 'Year 3X's Favourite Songs'. Type the five songs in the categories column and fill in the tally for each. Ask, 'Does this table represent our whole class?' Draw students' attention to the leftover sticky notes. Have students whose sticky notes were not included select a new 'favourite' song from the five songs on the list. Add these to the tally. Ask, 'Does this represent our whole class now?' Discuss.

WHOLE-CLASS INTRODUCTION

ENVIRONMENTAL AUDIT – DEVELOPING A RESEARCH QUESTION

You will need: NTO 3.13 'Data Collection Table'

Have students brainstorm ideas about an environmental issue in the school that is of relevance to them. Decide on an issue, such as litter, and develop ideas through questioning, e.g. 'What type of litter is in the school ground? Which area of the school ground has the most litter? How much rubbish do we have each day in the classroom? What types of rubbish do we have in the classroom?' Once an issue is selected, discuss and decide on categories, e.g. plastic, paper and compost. Ensure categories are clear and do not overlap to avoid confusion and incorrect data collection. Display NTO 3.13 'Data Collection Table' on the IWB and fill in the title and categories.

INDEPENDENT TASKS

Note: Choose from Tasks 1, 2 or 3.

You will need: BLM 29 'Data Collection Table', *Excel*, Student Book p. 44 'Collecting Data'

TASK 1: ENVIRONMENTAL AUDIT – COLLECTING DATA

Give students BLM 29 'Data Collection Table' and have them fill in the title and categories from the Whole-Class Introduction. Have them collect relevant data and fill in the table. This information will be used throughout the unit.

TASK 2: INTERACTIVE TASK

Have students use the Births, Deaths and Marriages Registry website for your state to look up the top ten names for their gender and year of birth. (Select the 'births' option then the 'popular names' link. Queensland and Western Australia do not have the option to select the year.) Have students use *Excel* to create a list of the top ten names in one column; in the next column, create a tally of the people they know who have this name; and in the next column, the total.

TASK 3: STUDENT BOOK p. 44 *'Collecting Data'*

TEACHING GROUP

You will need: sticky notes, NTO 3.13 'Data Collection Table'

MY PETS

- For students who require support, give them some sticky notes and have them write each pet they own on a separate sticky note. Ask, 'What are the main kinds of pets we have?' On NTO 3.13 'Data Collection Table', type the categories of pets, e.g. dogs, cats, fish. Once five or six types of pets are listed, type 'other' as the final category. Explain that we use 'other' so that we don't have to list every possible option. Have students enter their sticky notes into the 'tally' column then find the totals.

COLLECTING DATA ABOUT OUR SCHOOL

- For students who require a challenge, ask them to think of another issue in the school on which they could collect data, e.g. the amount of fruit scraps that are thrown away each day. Have them develop a method for collecting the data. If possible, allow them to implement their plan to check its effectiveness.

REFLECTION

Select from the following to suit your class and their learning outcomes:

- Ask, 'Why is it important that we organise the data we are collecting into categories?'
- Ask, 'What might happen if the categories we are using to collect data are not clear?'

LESSON PLAN 2

TUNING IN

HUMAN GRAPHS

You will need: five sheets of A4 paper, digital camera

Before the lesson, take five sheets of paper; write a colour on one side and a fruit on the other so you have five colours and five fruits. Place these (colours facing up) at the front of the room. Have students sit in a line behind the colour they like best. Ensure their lines are straight and evenly spaced. Discuss the number of people in each line, and how this forms a human graph. Take a photograph. Repeat using the fruit names.

WHOLE-CLASS INTRODUCTION

ENVIRONMENTAL AUDIT – DISPLAYING DATA

You will need: student data collection sheets from Lesson Plan 1, NTO 3.14 'Column Graph', *Word*

Display and discuss findings from the data collected in the environmental audit in Lesson Plan 1, Independent Tasks, Task 1. Explain that once data is collected, we need to display it to be easily understood. Show NTO 3.14 'Column Graph' on the IWB and ask, 'What is this? How could we use it to display our data?' Discuss. Ask, 'What would the title of our graph be?' Have students offer suggestions and add the title. Say, 'A column graph has two axes – vertical and horizontal. What are they for?' Discuss. (The horizontal axis holds the categories and the vertical axis shows us 'how many'.) Fill in the categories and the axes labels. Discuss the scale required on the vertical axis by looking at the highest number in the data table and deciding whether the data will fit – if not, count by 2s. Select students to add columns to the graph to represent the data. Save the graph by pressing Alt + PrtSc (this will copy the screen) then pasting it into *Word*.

INDEPENDENT TASKS

Note: Choose from Tasks 1, 2 or 3.

You will need: BLM 30 'Graph It!', *Excel*, Student Book p. 45 'Creating a Picture Graph'

TASK 1: GRAPHING DATA

Give students BLM 30 'Graph It!' and have them create their own graph of the environmental audit data collected in Lesson Plan 1. Ensure that the graph has a title, axes labels, categories and numbers/scale.

TASK 2: INTERACTIVE TASK 

Have students work independently on computers using *Excel* and the environmental audit data from Lesson Plan 1. In one column, they type the category names and in the next, the total for each category. Students then highlight the data, select the 'insert' tab and experiment with the different graphing features offered by this program.

TASK 3: STUDENT BOOK p. 45 *'Creating a Picture Graph'*

TEACHING GROUP

You will need: student data collection sheets from Lesson Plan 1, BLM 30 'Graph It!', counters, data about your school's class sizes separated by gender, BLM 10 'Grid Paper'

USING COUNTERS

- For students who require support, work as a group to create a graph of the class environmental audit data from Lesson Plan 1. Use an enlarged copy of BLM 30 'Graph It!', discuss and write in a title and labels, then have students place counters in each column to represent the data. Give students BLM 30 'Graph It!' and have them transfer the information from the group graph to their own page.

MORE COMPLEX GRAPHING

- For students who require a challenge, have them use the school class-size data and pose the question, 'How could we create a graph that shows how many boys and girls are in each class at our school?' Discuss options. Students use BLM 10 'Grid Paper' to create a column graph with two columns for each class (girls and boys) and a key to show the colours used for girls and boys.

REFLECTION

Select from the following to suit your class and their learning outcomes:

- Ask, 'How do graphs help us to understand our data?'
- Have students who completed the Student Book page show and explain their picture graphs to the class.

TUNING IN

INTERPRETING A PICTURE GRAPH

You will need: NTO 3.15 'Picture Graph'

Present NTO 3.15 'Picture Graph' to the class. Ask, 'This graph is about a Year 3 class. What could it be about? How many students might there be in the class?' Discuss, then ask, 'What if I said there are 30 students in the class – how could the data represent this?' Discuss so that students realise that each picture may be worth more than one. Liken this to counting by 2s on the vertical axis.

WHOLE-CLASS INTRODUCTION

ENVIRONMENTAL AUDIT – INTERPRETING DATA

You will need: NTO 3.16 'Audit Process'

Display NTO 3.16 'Audit Process' and track and discuss the progress so far. When you get to 'Interpret Data', ask, 'What do you think this means? What do we need to do now?' Discuss that the next step is to examine the data and write findings, i.e. what the data tells you. Display the saved graph from the Whole-Class Introduction in Lesson Plan 2 on the IWB. Discuss the graph, have students make factual statements about what the graph shows. Have each student write down three facts from the data. The next step, 'Drawing Conclusions' is the opportunity for students to ask 'why' and give reasons.

INDEPENDENT TASKS

Note: Choose from Tasks 1, 2 or 3.

You will need: A3 paper, NTO 3.15 'Picture Graph', Student Book p. 46 'Interpreting Data'

TASK 1: PUTTING IT ALL TOGETHER

Have students create a poster of their environmental audit and include a data collection table, a graph, three findings and reasons for their findings, as well as suggestions for how the sustainability of the school could be improved.

TASK 2: INTERACTIVE TASK

Have students experiment with NTO 3.15 'Picture Graph' to create graphs of their own. Have them swap graphs with another student, and write down three findings from each other's graphs.

TASK 3: STUDENT BOOK p. 46 *'Interpreting Data'*

TEACHING GROUP

You will need: environmental audit graph from Whole-Class Introduction, Lesson Plan 2; newspapers

INTERPRETING GRAPHS AS A GROUP

- For students who require support, continue to examine the class graph. Examine the data as a group and discuss findings. Encourage comparisons between columns using 'more' or 'less' statements.

GRAPHS IN NEWSPAPERS

- For students who require a challenge, give them a couple of different newspapers and have them each search for one or two graphs. Students share the graphs they have found with the group, then each student selects one graph (it does not have to be their own) to report at least three findings.

REFLECTION

Select from the following to suit your class and their learning outcomes:

- Have students present their posters from Independent Tasks, Task 1, and share suggestions to improve sustainability.
- Ask, 'Can you think of any real-life examples of when you need to interpret graphs?'

Home Tasks

Select from the possible Home Tasks:

- Have students think of an audit they could do at home, e.g. cutlery or DVD audit, and follow same process used in class (shown on NTO 3.16 'Audit Process').
- Have students discuss the process and results of the environmental audit with their parents or carers.

Assessment

- Have students complete **Student Assessment p. 47**.
- Review with students **Assessment Task Card 3.11**.

During the three lessons:

- Collect/make copies of students' data collection tables, graphs and interpretations to add to their portfolios as evidence of their ability to collect, use and interpret data.
- Make digital copies of *Excel* spreadsheets and graphs to add to students' digital portfolios.
- Collect written reflections on the environmental data audit as evidence of learning about how to use and interpret data, as well as heightening environmental awareness.

Recommendations for Future Learning

Specific to Student Assessment p. 47; if the student is experiencing some difficulty:

Q 1 Revise what a data collection table is and how to use it. Revisit the work on using a tally to sort data in Lesson Plan 1, Tuning In. If the student finds using the tally difficult, counters can be used to show the amounts in the total column.

Q 2 Revisit the work on displaying data in Lesson Plan 2. Remind the student what a column graph is and how it is labelled. Show the student a sample of a column graph if required.

Q 3 Revisit the interpreting data activities from Lesson Plan 3, reminding the student that findings need to be facts that are shown in the graph. Have the student explain what the graph is showing to ensure they have a clear understanding. Encourage comparison between columns using 'more' or 'less' statements.

If the student has not achieved the recommended skills for this unit:

1. See **Assessment Task Card 3.11** for specific recommendations.
2. Review the basic construction of a graph, e.g. axes, labels, title.
3. Have the student work with graphs using modelling equipment, then move to picture, then column graphs.
4. Review *Nelson Maths: Australian Curriculum NSW Year 2* Unit 22.

If the student has achieved the recommended skills and these skills are firmly established, consider:

1. Moving forward to *Nelson Maths: Australian Curriculum NSW Year 4* Units 20, 29 and 30.
2. Extending the student to work with larger values and more data amounts.
3. Having the student develop the use of spreadsheet applications.

Unit 12 Mental Strategies for Subtraction

Number and Algebra
Addition and subtraction MA2-5NA uses mental and written strategies for addition and subtraction involving two-, three-, four- and five-digit numbers

100 chart, count down, count up, counting, difference, digit, number line, strategy, subtract, take away

LESSON PLAN 1

TUNING IN

SUBTRACTION RACE

You will need: BLM 15 'Number Cards 1–20'

Give each student a card made from BLM 15 'Number Cards 1–20' and have them stand in a line at one end of the room. Say a subtraction fact that has an answer between 1 and 20. Have students mentally solve the subtraction fact and if they are holding the card that is the answer for that equation, they take one step forward. Continue with other subtraction facts. The winner is the student who reaches the other end of the room first. Discuss the strategies students used to solve the subtraction problems.

WHOLE-CLASS INTRODUCTION

SUBTRACTION STRATEGIES

You will need: poster paper

Present students with subtraction facts, e.g. 15 – 4, 24 – 6, 7 – 2, 26 – 15. Ask, 'What strategies could we use to solve these subtraction facts?' Discuss students' ideas and have them model how they solve a fact using their strategy. Make a poster of the familiar strategies, such as taking away with materials, counting back, and counting on to find the difference. Students refer to the poster when solving subtraction problems.

INDEPENDENT TASKS

Note: Choose from Tasks 1, 2 or 3.

You will need: BLM 14 '100 Chart', BLM 15 'Number Cards 1–20', counters, NTO 3.9 '100 Chart', BLM 1 '2-Digit Number Cards', dice, Student Book p. 48 'Ready, Set, Subtract!'

TASK 1: SUBTRACT TO REACH 100

Give pairs of students BLM 14 '100 Chart' and cards made from BLM 15 'Number Cards 1–20'. Students use the 100 chart as a game board, beginning by each placing a counter before number 1. They place the pile of cards upside down between them, and take turns to turn over two cards and make a subtraction problem using the two numbers. When they have worked out the answer, they move their counter that number of spaces. For example, for 15 – 8 = 7, they move their counter seven spaces. The winner is the first to reach 100.

TASK 2: INTERACTIVE TASK 

Give pairs of students cards made from BLM 1 '2-Digit Number Cards' and a dice. They take turns to turn over a card and roll the dice, then form a subtraction problem using these two numbers, e.g. 67 – 5. Students record and solve their problem, then highlight the answer on NTO 3.9 '100 Chart' (with 'show numbers' selected). The aim is to highlight as many squares as they can on the 100 chart.

TASK 3: STUDENT BOOK p. 48 *'Ready, Set, Subtract!'*

TEACHING GROUP

You will need: BLM 15 'Number Cards 1–20', BLM 1 '2-Digit Number Cards'

SUBTRACTION FLIP

- For students who require support, have them work with a partner and give them a set of cards made from BLM 15 'Number Cards 1–20'. Have them each turn over two cards and then each form a subtraction problem using the numbers on the cards. Students record their subtraction problem, solve it and record the strategy they used. The student with the smallest answer wins a point. The game continues with each student turning over another two cards.

SUBTRACTING 9

- For students who require a challenge, present them with cards made from BLM 1 '2-Digit Number Cards'. Select a card and ask, 'What would be the quickest way to subtract 9 from this number?' Discuss and model the strategy of subtracting 10 and adding 1. Ask, 'Why is this a faster way of solving the problem?' Have students practise this strategy by selecting cards and subtracting 9.

REFLECTION

Select from the following to suit your class and their learning outcomes:

- Have students share their work from the Independent Tasks. Ask, 'What strategies did you use when solving subtraction problems? How do you know what strategy is the most effective one to use?'
- Have students reflect on the 'Subtracting 9' activity from the Teaching Groups. Ask, 'What type of problems will this strategy help you solve?'

LESSON PLAN 2

TUNING IN

WHAT DO YOU NOTICE?

You will need: NTO 3.7 'Modelling with MAB'

Record 16 – 4 = 12 and model it using NTO 3.7 'Modelling with MAB'. Repeat for 26 – 4 = 22 and 36 – 4 = 32. Ask, 'What do you notice about these subtraction problems?' Record 46 – 4. Have students predict the answer. Ask, 'What strategy did you use?' Brainstorm other problems that can be solved using this pattern.

WHOLE-CLASS INTRODUCTION

SERIAL SUBTRACTION

You will need: NTO 3.9 '100 Chart'

Present the equation: 28 – 5. Solve the problem using NTO 3.9 '100 Chart' (select 'show numbers') by identifying the number 28 and then counting back five spaces. Say, 'We know that 28 – 5 = 23. How could that help us solve these equations?' Present the equations: 48 – 5, 58 – 5, 68 – 5. Have students solve the problems using the 100 chart. Discuss what students notice. Present the equation: 38 – 15. Ask, 'How could we use the pattern to help us solve this problem?'

INDEPENDENT TASKS

Note: Choose from Tasks 1, 2 or 3.

You will need: BLM 14 '100 Chart', dice, NTO 3.7 'Modelling with MAB', Student Book, p. 49 'Serial Subtraction'

TASK 1: HIGHLIGHT THE PATTERN

Present students with the following problems: 47 – 5 = 42 and 37 – 5 = 32. Give them BLM 14 '100 Chart' and have them highlight the two subtraction problems on the chart. Have them highlight and record other subtraction problems that could be solved using the pattern. Repeat with the following equations: 36 – 2 and 46 – 2; 77 – 3 and 67 – 3.

TASK 2: INTERACTIVE TASK

Have students work with a partner, taking turns to roll three dice and then making a 2-digit number and a 1-digit number. For example, if they roll 3, 5 and 4, they could make 35 and 4. Have students form a subtraction problem using their numbers, e.g. 35 – 4, and model it using NTO 3.7 'Modelling with MAB'. They then take turns to continue the serial subtraction pattern by modelling and recording different subtraction problems, such as 45 – 4 or 65 – 4.

TASK 3: STUDENT BOOK p. 49 *'Serial Subtraction'*

TEACHING GROUP

You will need: NTO 3.6 'Number Line'

NUMBER LINES

- For students who require support, present them with the equation 35 – 4. Guide them to solve the problem using NTO 3.6 'Number Line', identifying the number 35 and then jumping back 4 numbers to stop at 31. Have students then solve the problems 45 – 4, 65 – 4 and 25 – 4. Ask, 'What can you see happening?' Repeat with other equations, such as 73 – 5, 63 – 5, 53 – 5 and 23 – 5. Ask, 'What other problems could we solve in this way?'

MORE SERIAL SUBTRACTION

- For students who require a challenge, present the equations: 67 – 5, 67 – 15, 77 – 25. Have them solve the problems and then discuss what they notice. Discuss how the number of ones being subtracted in each problem is the same, while the number of tens being subtracted changes. Ask, 'How did the equation 67 – 5 = 62 help you solve the other equations?' Repeat with the equations 58 – 7, 68 – 17 and 78 – 37, and 49 – 6, 39 – 16 and 79 – 26. Have students then record other 2-digit subtraction problems that can be solved using the serial subtraction strategy.

REFLECTION

Select from the following to suit your class and their learning outcomes:

- Have students explain the patterns they noticed when solving serial subtractions.
- Ask, 'If one of your friends was having problems solving serial subtraction, how could you help?'

LESSON PLAN 3

TUNING IN

SUBTRACTING TENS

You will need: BLM 1 '2-Digit Number Cards', NTO 3.7 'Modelling with MAB'

Give students cards made from BLM 1 '2-Digit Number Cards', placed face down in the middle of the group. Select a student to turn over a card. Ask, 'How can you take away 10 from this number?' Have the student model the number using NTO 3.7 'Modelling with MAB'. Discuss how '– 10' is the same as taking away a group of ten. Have students remove a ten from the modelled number. Have a student turn over another card and ask, 'How could you take away 20 from this number?' When they have modelled their number, guide them in taking away two tens. Continue, subtracting 40, 50 or 80. Ask, 'How could this strategy help you to solve 63 – 24?' Talk about how students could identify and subtract the tens, then the ones, i.e. 63 – 20 – 4.

WHOLE-CLASS INTRODUCTION

SOLVE IT ON A NUMBER LINE

You will need: NTO 3.6 'Number Line'

Present the equation: 54 – 20. Ask, 'How could we solve this equation on a number line?' Use NTO 3.6 'Number Line' to model beginning at 54 and jumping back 10 to get to 44 and then jumping back another 10 to get to 34. Present the problem: 73 – 30. Ask, 'How many tens would we need to jump back to solve this problem?' Select a student to model jumping back 3 tens to stop at 43. Ask, 'How could we use this to solve 56 – 21?'

INDEPENDENT TASKS

Note: Choose from Tasks 1, 2 or 3.

You will need: NTO 3.10 'Card Flip', BLM 31 'Counting by 10 Cards', BLM 32 'Number Lines', dice, small whiteboards, NTO 3.11 'Calculator', Student Book p. 50 'Orchard Farmer'

TASK 1: NUMBER LINES

Have students make subtraction problems by turning over a 0–99 card on NTO 3.10 'Card Flip' and a card from BLM 31 'Counting by 10 Cards'. For example, if the number 63 was shown on the NTO and they turned over the 20 card, the equation would be 63 – 20. Have students solve the equations using number lines on BLM 32 'Number Lines'. For example, for 63 – 20, students would start at 63 on the number line, then jump back to 53 to take away 10 and then jump back another 10 to 43, which would be the answer.

TASK 2: INTERACTIVE TASK 

Have the students roll two dice to make a 2-digit number, e.g. if they roll a 4 and 6 they could make 46. Have students subtract 10 from that number and record the answer on their whiteboard. Students can then enter the equation on NTO 3.11 'Calculator'. When they have the answer, they then take away 10 from that number, i.e. 36 – 10. Students continue to take away 10 from each answer until they reach a 1-digit number. Each time, they are to predict what the answer will be on their whiteboard before they solve it using the calculator. Guide students in recognising that as they take away 10, the digit in the tens column changes.

TASK 3: STUDENT BOOK p. 50 *'Orchard Farmer'*

TEACHING GROUP

You will need: BLM 31 'Counting by 10 Cards', MAB

MODELLING TENS

- For students who require support, present them with cards made from BLM 31 'Counting by 10 Cards'. Have them select two cards and make a subtraction equation using those numbers. Reinforce how the larger number is first in subtraction equations. Have students model the larger number using MAB, then ask, 'How many tens do you need to take away?' Guide students to take away the appropriate number of tens and ask them to explain how they worked out the answer.

USE THE STRATEGY

- For students who require a challenge, present the equation: 46 – 21. Ask, 'How could we use the subtracting tens strategy to help solve this problem?' Discuss how 21 is one more than 20. Model subtracting 20 and then subtracting another 1. Present the equation: 46 – 19. Discuss how 19 is one less than 20. Model subtracting 20 and then adding 1. Ask, 'Why is this a good strategy to use?' Have students solve the following equations using this strategy: 58 – 21, 76 – 31, 84 – 39 and 64 – 29.

REFLECTION

Select from the following to suit your class and their learning outcomes:

- Have students reflect on Independent Tasks, Task 2. Ask, 'What did you notice as you continually subtracted 10 from a number? Why was the number in the tens column changing?'
- Encourage students to connect their understanding of subtracting tens to their knowledge of adding tens. Ask, 'How are these strategies similar? How are they different?'

Home Tasks

Select from the possible Home Tasks:

- Have students find two numbers in a newspaper (electronic or hard copy) and form and solve a subtraction problem using these numbers. Discuss the strategies they use to solve the problems.
- Have students identify 2-digit numbers around the house, e.g. in the pantry, in books, on the TV or in catalogues. Have them subtract 10, 30 or 50 from these numbers.

Assessment

- Have students complete **Student Assessment p. 51**.
- Review with students **Assessment Task Card 3.12**.

During the three lessons:

- Observe which students are able to use mental strategies to solve the subtraction equations.
- Make note of students who completed the scaffolding tasks or the more challenging activities of the Teaching Groups.
- Review Student Book pages and make notes of areas of difficulty.

Recommendations for Future Learning

Specific to Student Assessment p. 51; if the student is experiencing some difficulty:

Q 1–2 Revise the strategies for solving subtraction problems, such as counting back, counting on, finding the difference and taking away. Present the student with simple subtraction equations, such as 9 – 5. Have the student use the strategies to solve the problems.

Q 3 Revise serial subtraction. Present equations: 9 – 6 = 3, 19 – 6 = 13 and 29 – 6 = 23. Have the student discuss what they notice. Have the student explore serial subtraction patterns, using number lines or a 100 chart.

Q 4–5 Have the student review the connection between '– 10' and subtracting a group of ten. Encourage the student to use MAB to model subtracting 10, 20, 30 and 50 from 2-digit numbers. Emphasise how the tens decrease when solving these subtraction problems.

If the student has not achieved the recommended skills for this unit:

1. See **Assessment Task Card 3.12** for specific recommendations.
2. Have the student revise strategies, such as counting back, for solving simple subtraction problems.
3. Provide opportunities for the student to practise and consolidate the mental subtraction strategies.
4. Review *Nelson Maths: Australian Curriculum NSW Year 2* Unit 11 and *Building Mental Strategies Skill Book Year 3*, pp. 48–54.

If the student has achieved the recommended skills and these skills are firmly established, consider:

1. Having the student complete *Building Mental Strategies Skill Book Year 4*, pp. 38–43, to reinforce mental subtraction strategies.
2. Moving forward to *Nelson Maths: Australian Curriculum NSW Year 4* Unit 13.
3. Extending the student by exploring mental subtraction strategies involving decimals and fractions.

Unit 13 Subtraction

Number and Algebra

Addition and subtraction MA2-5NA uses mental and written strategies for addition and subtraction involving two-, three-, four- and five-digit numbers

difference, estimating, nearest 10, ones, renaming, subtract, take away, tens,

LESSON PLAN 1

TUNING IN

BRAINSTORM

Have students brainstorm things they know about subtraction. Guide them to discuss concepts and strategies, such as the bigger number being first in the equation, counting back and the answer being smaller than the number it is being taken away from. Record ideas and display in the room for reference.

WHOLE-CLASS INTRODUCTION

MODELLING SUBTRACTION WITH MAB

You will need: NTO 3.7 'Modelling with MAB'

Present the equation: 46 – 25. Have students discuss what they need to do to solve this equation. Ensure they understand that they begin with 46 and need to take 25 away. Model 46 using NTO 3.7 'Modelling with MAB'. Discuss how 25 does not need to be modelled because it is the amount being taken away. Model taking away the 5 ones and 2 tens to calculate the answer of 21. Present the equation: 53 – 36. Model 53 and establish the need to take away 6 ones and 3 tens. Ask, 'How can we take away 6 ones if there are only 3 ones in 53?' Model renaming 1 ten as 10 ones. Discuss how there are now 4 tens and 13 ones. Model taking away the 6 ones and 3 tens to have 17 remaining. Provide more problems, such as 58 – 24 and 62 – 37.

INDEPENDENT TASKS

Note: Choose from Tasks 1, 2 or 3.

You will need: BLM 8 'Place-Value Mat', 10-sided dice, MAB, BLM 33 'Subtraction Cards', NTO 3.7 'Modelling with MAB', Student Book p. 52 'At the Fair'

TASK 1: RACE TO ZERO

Give pairs of students BLM 8 'Place-Value Mat', a 10-sided dice and MAB. The game begins with students modelling 100 on the place-value mat using MAB. They take turns to roll the dice and subtract the number shown on the dice from the MAB. For example, if a student rolls a 6, they would take away 6 from the ones column on the mat. If there are not enough ones available, students need to rename 1 ten as 10 ones. For example, if they need to subtract 8 ones but there are only 2 ones on the mat, they take 1 ten from the tens column, swap it for 10 ones and place the ones in the ones column on the mat. This would make 12 ones and the student could then subtract 8. The student who subtracts the last amount so there is zero on the mat is the winner.

TASK 2: INTERACTIVE TASK 

Give students cards made from BLM 33 'Subtraction Cards'. Have students model these equations using NTO 3.7 'Modelling with MAB'. Support them as they subtract the ones and tens, and rename tens for ones as needed.

TASK 3: STUDENT BOOK p. 52 *'At the Fair'*

TEACHING GROUP

You will need: MAB, dominoes

SUBTRACTION WITH NO RENAMING

- For students who require support, revise subtraction equations that do not involve renaming, such as 68 – 42. Have students model 68 using MAB. Ask, 'How many ones and how many tens do we need to take away?' Emphasise the importance of subtracting the ones first and then the tens. Have students solve: 57 – 25, 69 – 32, 74 – 13, 88 – 35 and 48 – 21.

DOMINO CHALLENGE

- For students who require a challenge, have them work with a partner to make subtraction equations using dominoes. Show students a domino and explain how it can be read, e.g. as 45 or turned around and read as 54. Have students place the dominoes face down and then select two dominoes each. They read the dominoes as 2-digit numbers and use them to form a subtraction problem. Both students solve their subtraction problem; the one with the smallest answer scores a point.

REFLECTION

Select from the following to suit your class and their learning outcomes:

- Ask, 'Why do you need to rename a ten as 10 ones when solving some subtraction problems?'
- Write the equation 57 – 35 on the board. Have students explain how they would solve this problem. Encourage them to use MAB and the board to support their explanation.

LESSON PLAN 2

TUNING IN

WHERE IS THE MISTAKE?

You will need: NTO 3.11 'Calculator'

Present the equation: 56 – 18 = 42. Ask, 'How could we work out if a mistake was made here?' Students brainstorm ways of checking subtraction problems, such as revising or using a calculator. Have students discuss how they would revise the problem. As a group, revise the problem, identify where the error was made and solve it correctly. Then check the answer using NTO 3.11 'Calculator'. Ask, 'Why do you think a mistake was made here?'

WHOLE-CLASS INTRODUCTION

VERTICAL RECORDING

Present the equation: 63 – 21 (horizontally). Explain that this equation can be recorded in another way. Record the equation vertically and discuss how the tens and ones in each number are lined up vertically. Emphasise that the ones need to be subtracted first and then the tens. Model how the answer is recorded, then present: 54 – 28 (horizontally). Have a student show how the problem can be recorded vertically. Ask, 'How can we subtract 8 ones when there are only 4 ones in 54?' Discuss how a ten can be renamed as 10 ones and model how this is recorded on the algorithm. Then subtract the tens and record the answer. Repeat with 56 – 21 and 73 – 26.

INDEPENDENT TASKS

Note: Choose from Tasks 1, 2 or 3.

You will need: BLM 34 'Bingo Boards', BLM 33 'Subtraction Cards', counters, small whiteboards, dice, NTO 3.11 'Calculator', Student Book p. 53 'Correction Time'

TASK 1: TAKE AWAY BINGO

Give pairs of students 'Bingo' boards from BLM 34 'Bingo Boards', a set of cards from BLM 33 'Subtraction Cards', counters and a small whiteboard. Have each student take a 'Bingo' board and place the subtraction cards face down in a pile. Students take turns to turn over a card. Have students record the problem vertically on the whiteboard, solve it and then look on their 'Bingo' board to see if they have the answer. If they do, they cover the answer square with a counter. The winner is the first student to cover all the squares.

TASK 2: INTERACTIVE TASK 

Have students work in pairs to roll four dice and make two 2-digit numbers and a subtraction problem. Have one student record the equation vertically on a small whiteboard and then solve it. Their partner solves the equation using NTO 3.11 'Calculator'. Students compare their answers to ensure they have solved it correctly. Students then swap roles.

TASK 3: STUDENT BOOK p. 53 *'Correction Time'*

TEACHING GROUP

You will need: MAB, poster paper, deck of playing cards (picture cards removed), sticky notes, BLM 1 '2-Digit Number Cards'

STEP BY STEP

- For students who require support, present the equations: 65 – 31, 73 – 28. Record each problem vertically and solve them using MAB. As a group, make a poster recording the steps taken. Emphasise how one equation required renaming (swapping 1 ten for 10 ones) and the other did not. Then give students a deck of cards and have them each turn over four cards to make two 2-digit numbers. Have students use their numbers to record a vertical subtraction problem on a sticky note. When they have solved the problems, students stick the equations that required renaming in one area and the equations that did not require renaming in another. Ask, 'What do you notice about the equations in each group?'

THE ANSWER IS ...

- For students who require a challenge, give them cards made from BLM 1 '2-Digit Number Cards'. Write the numbers 56, 27, 14, 20, 28, 48 and 25 on the board. Explain that these numbers are the answers to subtraction problems. Have students use the cards to make subtraction equations that have the numbers on the board as the answer. Discuss the strategies students used.

REFLECTION

Select from the following to suit your class and their learning outcomes:

- Have students explain things to remember when recording and solving subtraction problems vertically.
- Have students discuss how they know if a subtraction problem requires renaming. Ask them to generate a subtraction problem that requires renaming and one that does not.

LESSON PLAN 3

TUNING IN

ROUNDING PREDICTION

You will need: NTO 3.10 'Card Flip'

Tell students that one side of the room is 'up' and the other side is 'down'. Explain that you are going to show a number and they need to predict if it will be rounded up or down to the nearest ten, by standing on the 'up' or 'down' side of the room. Present a number using NTO 3.10 'Card Flip', selecting 0–99 cards. If it would be rounded up, e.g. 47, students on the 'down' side are out. If it would be rounded down, e.g. 42, students on the 'up' side are out. Those still in the game make a new prediction before the next number is shown. The game continues until one student is left, who is the winner.

WHOLE-CLASS INTRODUCTION

COMPETITION

You will need: BLM 33 'Subtraction Cards', BLM 31 'Counting by 10 Cards'

Discuss how rounding numbers can help with solving or checking answers to subtraction problems. Present the equation: 57 – 21. Ask, 'How could rounding numbers help us solve this equation?' Explore steps including rounding each of the numbers up or down and then subtracting these numbers. Give each student a card from BLM 31 'Counting by 10 Cards' and a problem from BLM 33 'Subtraction Cards'. Have them estimate the answer by rounding the numbers to the nearest ten and then subtracting those numbers. If the estimate is the same number that is on their card, the student receives 10 points. The winner has the most points at the end.

INDEPENDENT TASKS

Note: Choose from Tasks 1, 2 or 3.

You will need: BLM 1 '2-Digit Number Cards', calculators, LO: *L100 'The take-away bar: generate hard subtractions'*, Student Book p. 54 'Match the Estimation'

TASK 1: ESTIMATE AND CHOOSE

Give students cards made from BLM 1 '2-Digit Number Cards' placed face down in a pile. Tell students that they will be trying to make a subtraction equation with a difference of 40. Have students each turn over two cards, use the numbers revealed to make their subtraction problem and estimate the difference between their two numbers. They can choose to keep both of their cards or return one card to the bottom of the pile and select another card from the top. Students can then use the calculators to solve their equation. The student with the answer closest to 40 wins a point. Students return their cards to the bottom of the pile and continue the game.

TASK 2: INTERACTIVE TASK 

Have students work on computers, using LO: *L100 'The take-away bar: generate hard subtractions'* to solve subtraction equations. Before entering their final answer, have students estimate the answer by rounding the numbers in the equation up or down to the nearest ten and subtracting these numbers. If their answer is not similar to their estimation, students should revise the equation.

TASK 3: STUDENT BOOK p. 54 ***'Match the Estimation'***

TEACHING GROUP

You will need: sticky notes, calculators, NTO 3.10 'Card Flip', BLM 1 '2-Digit Number Cards'

STICKY PROBLEMS

- For students who require support, have them generate and record their own 2-digit subtraction problems on sticky notes. Guide them as they round their numbers to the nearest ten and then subtract these numbers to estimate the answer to their equation. As a group, sort the subtraction problems by sticking together those with an estimated answer of 10, 20, 30 and so on. Have students use calculators to see if their estimations were accurate.

ESTIMATING BIGGER NUMBERS

- For students who require a challenge, have them estimate with larger numbers. Present students with NTO 3.10 'Card Flip', selecting 100–999 cards. As a number is shown, have them simultaneously flip over a card from BLM 1 '2-Digit Number Cards'. Students form a subtraction equation with these two numbers. Ask them to estimate the answer by rounding the numbers to the nearest ten and then solving that subtraction problem. Students can use a calculator to see if their estimation was accurate. Repeat by turning over other cards.

REFLECTION

Select from the following to suit your class and their learning outcomes:

- Have students share their work from the Independent Tasks or the Teaching Groups and discuss how they used the estimation strategy. Ask, 'How can estimation and rounding help you solve subtraction problems?'
- Discuss what students would do if their answer and estimation were not similar.

Home Tasks

Select from the possible Home Tasks:

- Have students find numbers in a newspaper and use them to make subtraction problems.
- Encourage students to help in cooking activities. Have them recognise and record subtraction equations during the process, e.g. 'We had 65 grams of butter and there is now 20 grams left. How much did we use?'

Assessment

- Have students complete **Student Assessment p. 55**.
- Review with students **Assessment Task Card 3.13**.

During the three lessons:

- Identify who is able to regroup and estimate accurately when completing subtraction problems.
- Review Student Book pages and make notes of areas of difficulty.

Recommendations for Future Learning

Specific to Student Assessment p. 55; if the student is experiencing some difficulty:

Q 1 Have the student solve 2-digit subtraction equations by modelling the process using MAB. Have the student also use MAB to model renaming 1 ten as 10 ones.

Q 2–3 Discuss the specific steps involved in solving subtraction equations, such as subtracting the ones, subtracting the tens and renaming. Have the student solve subtraction problems by modelling with MAB. Encourage the student to think aloud as they solve the problems to ensure they are following all the steps.

Q 4–5 Have the student revisit how they estimate the answers to subtraction problems and practise rounding the 2-digit numbers in the equations to the nearest ten and subtracting these numbers. Discuss how estimating is a way of checking the reasonableness of their answer to a subtraction problem.

If the student has not achieved the recommended skills for this unit:

1. See **Assessment Task Card 3.13** for specific recommendations.
2. Have the student subtract 2-digit with 2-digit numbers without renaming before solving subtraction problems involving renaming.
3. Review *Nelson Maths: Australian Curriculum NSW Year 2* Unit 12.

If the student has achieved the recommended skills and these skills are firmly established, consider:

1. Having the student complete *Building Mental Strategies Skill Book Year 4*, pp. 40–43, to reinforce mental strategies for subtraction.
2. Moving forward to *Nelson Maths: Australian Curriculum NSW Year 4* Unit 13.
3. Extending the student in any of the listed activities by using larger numbers or numbers involving decimals.

Connections Between Addition and Subtraction

Number and Algebra

Addition and subtraction MA2-5NA uses mental and written strategies for addition and subtraction involving two-, three-, four- and five-digit numbers

addition, balanced, equal, fact family, plus, related problem, subtraction, take away, ten frame, unbalanced

LESSON PLAN

TUNING IN

TEN FRAME ADDITION AND SUBTRACTION

You will need: NTO 3.1 'Ten Frames'

Present the equation: 4 + 6 = 10. Use NTO 3.1 'Ten Frames' to model the equation, selecting four red counters and six blue counters. Discuss how when the two amounts are added together, it makes a total of 10. Rearrange the counters in the ten frame to model 6 + 4 = 10. Record and discuss how the two addends have been rearranged. Ask, 'Why are the answers to these two addition problems the same?' Then model and record 10 – 6 = 4. Emphasise how the four red counters remain when the six blue counters are taken away. Then model and record 10 – 4 = 6. Ask, 'What did you notice about these addition and subtraction problems?'

WHOLE-CLASS INTRODUCTION

FIND YOUR FAMILY

You will need: BLM 15 'Number Cards 1–20', small whiteboards

Write 3, 5 and 8 on the board. Ask, 'What addition and subtraction problems can we make using these numbers?' Record 3 + 5 = 8, 5 + 3 = 8, 8 – 5 = 3 and 8 – 3 = 5 on the board. Discuss how these numbers form a fact family because the same numbers are used in all four equations. Have students identify the parts (3 and 5) and the whole (8) in each equation. Explain that a fact family can be made with any addition or subtraction problem. Give each student a card from BLM 15 'Number Cards 1–20' and have them move around the room and form fact family groups of three students whose number cards can be used together to form a fact family, e.g. 5, 7 and 12. Then have students form and record the two addition problems and the two subtraction problems from their fact family group on the whiteboards.

INDEPENDENT TASKS

Note: Choose from Tasks 1, 2 or 3.

You will need: BLM 15 'Number Cards 1–20', blank squares of paper, coloured pencils, NTO 3.17 'Ten-Sided Dice', Student Book p. 56 'Fact Families'

TASK 1: REARRANGE THE CARDS

Give each student a set of cards made from BLM 15 'Number Cards 1–20' and three blank squares of paper. Have them draw an addition symbol, a subtraction symbol and an equals symbol on their squares and then form an addition or subtraction problem using the number cards and symbols, e.g. 3 + 4 = 7. Have students record their equation, then rearrange the cards to form the related addition and subtraction equations.

TASK 2: INTERACTIVE TASK

Present students with NTO 3.17 'Ten-Sided Dice'. As a number is shown, have students record an addition or subtraction equation using that number, e.g. if 8 is shown, students may form 8 + 3 = 11. Students then record the related facts to that equation, e.g. 3 + 8 = 11, 11 – 3 = 8 and 11 – 8 = 3.

TASK 3: STUDENT BOOK p. 56 *'Fact Families'*

TEACHING GROUP

You will need: BLM 35 'Fact Family Houses', counters

FACT FAMILY HOUSES

- For students who require support, give them BLM 35 'Fact Family Houses'. Identify the house with the numbers 2, 6 and 8. Use counters to model 2 and 6. Discuss how these can be put together to make 8.

Ask, 'Why do you think 2 and 6 are the walls of the house and 8 is on the roof?' Have students use the counters to model and record the two addition and two subtraction problems that relate to the fact family house. Emphasise how the 'whole' amount is at the end of the equation for addition and at the beginning for subtraction. Have students model and record equations using the other fact family houses.

MISSING NUMBER

- For students who require a challenge, have them work with a partner using BLM 35 'Fact Family Houses'. Have one student cover a number on a house with a counter. Their partner then tries to work out what the covered number is and records the possible equations. When the student thinks they have worked out the missing number, the counter is removed to check if they were correct. Students then swap roles and use a different fact family house. Discuss the strategies students used to work out the missing number.

REFLECTION

Select from the following to suit your class and their learning outcomes:

- Have students share their work from Independent Tasks, Task 1. Ask, 'How did you know the addition and subtraction problems were related? What did you notice about where the numbers were?'
- Present the equation: 15 – 11 = 4. Ask, 'What addition and subtraction problems would be in a fact family with this equation? What strategies would you use to work out the other equations?'

LESSON PLAN 2

TUNING IN

IS IT CORRECT?

Present the equation: 15 + 8 = 23. Ask, 'How could we check if this equation is correct?' Discuss strategies such as revising, estimating and using a calculator. Ask, 'How could we use our knowledge of the connection between addition and subtraction or fact families?' Write 23 – 8. Explain that if this has the answer of 15, we know our initial equation is correct. Have students solve the equation and explain why they know the addition problem is correct. Discuss why we are able to use addition to check subtraction problems or subtraction to check addition problems.

WHOLE-CLASS INTRODUCTION

MODEL THE NUMBERS

You will need: NTO 3.7 'Modelling with MAB'

Write 34 + 21 on the board. As a group, solve the addition problem by modelling and grouping together the two amounts using NTO 3.7 'Modelling with MAB'. Ask, 'How will we know if 45 is the correct answer?' Discuss how solving the subtraction problem 45 – 21 will prove if the answer is correct. Take away 21 from 45 on the IWB and have students identify the answer. Discuss what they have noticed.

INDEPENDENT TASKS

Note: Choose from Tasks 1, 2 or 3.

You will need: dice, BLM 33 'Subtraction Cards', NTO 3.11 'Calculator', Student Book p. 57 'Related Addition and Subtraction'

TASK 1: PARTNER CHECK

Have students work with a partner, rolling four dice to form two 2-digit numbers. Have one student form and solve an addition problem using the two numbers. Their partner checks if the answer is correct by forming and solving a related subtraction problem. Students swap roles and roll the dice again.

TASK 2: INTERACTIVE TASK 

Give students cards made from BLM 33 'Subtraction Cards'. Have them solve the subtraction problems and then use NTO 3.11 'Calculator' to check if their answers are correct. Tell students that they may only enter addition problems into the calculator, so they will need to form the related addition problems to check their answers.

TASK 3: STUDENT BOOK p. 57 ***'Related Addition and Subtraction'***

TEACHING GROUP

You will need: BLM 3 'Blank Chart', BLM 1 '2-Digit Number Cards', NTO 3.7 'Modelling with MAB'

CUT UP 100

- For students who require support, use BLM 3 'Blank Chart'. Discuss how there are 100 squares on the chart. Cut the chart in half to make two pieces with 50 squares on each one. Ask, 'What addition problem can we make to represent this?' Record the equation 50 + 50 = 100. Move away one of the pieces and ask, 'What subtraction problem would match this situation?' Record 100 – 50 = 50. Discuss how the addition and subtraction problems are using the same numbers. Have students cut up copies of the blank chart to model other related addition and subtraction problems, e.g. 80 + 20 = 100 and 100 – 80 = 20.

PREDICTING

- For students who require a challenge, give them cards from BLM 1 '2-Digit Number Cards'. Select 24 and 45. Have students use NTO 3.7 'Modelling with MAB' to model and add the numbers to get 69. Have students subtract 24. Ask, 'Who can predict the answer?' Discuss the strategies students used. Model the subtraction using the IWB and ask, 'How did you know the answer was going to be 45?' Have students predict the answer if 45 was subtracted from 69. Pairs select two new cards, and repeat the activity.

REFLECTION

Select from the following to suit your class and their learning outcomes:

- Present the equation: 46 + 23 = 69. Ask, 'How could you use subtraction to check the answer to this?'
- Discuss what students should do if the subtraction problem they used to check an addition problem didn't have the same numbers as the addition problem. Ask, 'What would this make you think?'

LESSON PLAN 3

TUNING IN

MISSING NUMBER

You will need: sticky notes, NTO 3.7 'Modelling with MAB'

Write 12 + 6 = 18 on the board, but cover the 6 with a sticky note. Ask, 'What is the missing number?' Discuss strategies, such as counting on from 12 to 18. Ask, 'How could we use our understanding about the connection between addition and subtraction?' Discuss how forming and solving the problem 18 – 12 = 6 provides the missing number. Model the problem using NTO 3.7 'Modelling with MAB'. Remove the sticky note to reveal the missing number. Repeat with 25 + 34 = 59, covering the 25.

WHOLE-CLASS INTRODUCTION

BALANCE IT OUT

You will need: strips of paper

Write = on the board. Ask, 'What does this symbol mean?' Discuss how the amounts on both sides of the equals symbol need to be the same. Write 4 + 4 = 7 + 1 on the board. Establish that the 4 + 4 on the left-hand side of the equals sign makes 8 and the 7 + 1 on the right-hand side also makes 8. Discuss how this equation is balanced because both sides equal the same amount. On strips of paper, write simple addition facts, e.g. 4 + 1, 3 + 5, 2 + 3 and 6 + 2. Have students hold their arms out like a set of balance scales and explain that they are going to be shown two equations. If the equations are not equal, they hold one arm up and one arm down. If the two equations are equal, they hold their arms straight because they are balanced. Present two equations at a time and have students move their arms up or down as appropriate. Write 7 + 8 = 10 + ? on the board. Ask, 'What would the missing number be?' Discuss students' ideas.

INDEPENDENT TASKS

Note: Choose from Tasks 1, 2 or 3.

You will need: BLM 36 'Missing Number Maze', NTO 3.11 'Calculator', BLM 1 '2-Digit Number Cards', small whiteboards, Student Book p. 58 'Balancing Equations'

TASK 1: MISSING NUMBER MAZE

Have students complete BLM 36 'Missing Number Maze'. Encourage them to calculate the missing number by solving the related subtraction equation.

TASK 2: INTERACTIVE TASK 

Give students cards made from BLM 1 '2-Digit Number Cards'. Have them work with a partner, turning over two cards and forming an addition or subtraction problem with a missing number. For example, 28 + ___ = 46 or 46 – ___ = 28. Students use NTO 3.11 'Calculator' to find the missing number by entering a related subtraction problem, such as 46 – 28 = 18. Have students record the related equations on the whiteboards.

TASK 3: STUDENT BOOK p. 58 *'Balancing Equations'*

TEACHING GROUP

You will need: BLM 5 'Digit Cards', squares of paper, BLM 1 '2-Digit Number Cards'

BALANCED EQUATIONS

- For students who require support, give them cards made from BLM 5 'Digit Cards'. Have them work with a partner to make balanced equations. Have students write =, + and – symbols each on a square of paper. Have one student make an equation and place it on one side of the equals symbol, e.g. 4 + 2. Their partner then makes an equation to place on the other side of the equals symbol, e.g. 9 – 3. Students need to ensure that the equation is balanced. Have students swap roles and make other balanced equations.

FORM THE PROBLEMS

- For students who require a challenge, have them work with a partner to make addition and subtraction problems. Give them cards made from BLM 1 '2-Digit Number Cards'. Students take turns to select a card and present it to their partner. Their partner forms an addition and a subtraction problem that equals the number. Encourage students to use their understanding of the connection between addition and subtraction.

REFLECTION

Select from the following to suit your class and their learning outcomes:

- Have students share the strategies they used to identify the missing numbers in Independent Tasks, Task 1. Ask, 'How did a related addition or subtraction problem help you find the missing number?'
- Have students explain the connection between addition and subtraction.

Home Tasks

Select from the possible Home Tasks:

- Have students make addition and subtraction problems using numbers found in the newspaper. Have them swap the numbers around to form other problems using the same numbers.
- Have students look at the numbers on car number plates while travelling in a car, adding the numbers on each number plate to find the total. Have students look for number plates that have the same total.

Assessment

- Have students complete **Student Assessment p. 59**.
- Review with students **Assessment Task Card 3.14**.

During the three lessons:

- Observe which students are able to form problems that are part of the same fact family, check the answer to their problem by solving a related addition or subtraction problem, identify the missing number in a given equation and balance equations by using the connection between addition and subtraction.
- Note who completed the scaffolding tasks or the more challenging activities of the Teaching Groups.
- Review Student Book pages and make notes of areas of difficulty.

Recommendations for Future Learning

Specific to Student Assessment p. 59; if the student is experiencing some difficulty:

Q 1 Have the student use counters to model and explore an addition and subtraction fact family. Emphasise how each of the problems has the same numbers but they are in different positions. Present the student with other fact family groups, e.g. 6, 2 and 8. Have the student model the problems in these fact families.

Q 2 Have the student solve simpler addition problems, e.g. 7 + 5, using counters to model the problem. Guide the student to form a related subtraction problem, e.g. 12 – 7. Establish that the answer to the related subtraction problem is the number 5 from the original addition problem. When the student is proficient, introduce 2-digit with 2-digit addition problems.

Q 3–4 Revise how the missing number of an addition problem can be solved by using the given numbers to form a related subtraction problem. Emphasise how the numbers are the same, but in different positions.

Q 5 Revise what balanced equations are. Emphasise that both sides of the equals symbol need to equal the same amount. Discuss how to identify the missing number, such as solving the equation without a missing number and then using this answer to help solve the missing number on the other side.

If the student has not achieved the recommended skills for this unit:

1. See **Assessment Task Card 3.14** for specific recommendations.
2. Have the student find the missing number in equations involving 1-digit numbers and then move onto activities with 2-digit numbers. Provide opportunities for the student to use related addition or subtraction problems to check their answers when solving 1-digit with 1-digit addition and subtraction problems.
3. Review *Nelson Maths: Australian Curriculum NSW Year 2* Unit 16.

If the student has achieved the recommended skills and these skills are firmly established, consider:

1. Having the student complete *Building Mental Strategies Skill Book Year 4*, pp. 24–25 and 40–41, to reinforce mental strategies with addition and subtraction.
2. Moving forward to *Nelson Maths: Australian Curriculum NSW Year 4* Unit 13.
3. Extending the student in any of the listed activities by using larger numbers or numbers involving decimals.

Unit 15 Solving Addition and Subtraction Problems

Number and Algebra

Addition and subtraction MA2-5NA uses mental and written strategies for addition and subtraction involving two-, three-, four- and five-digit numbers

addition, difference, digit, regrouping, subtraction, take away, total, worded problem

LESSON PLAN 1

TUNING IN

WHAT'S OUR TOTAL?

You will need: sticky notes, NTO 3.11 'Calculator'

Have students sit in a circle, count each student and have them write their number on a sticky note. When all students have a number, ask, 'If we added all our numbers together, what would our total be?' Discuss strategies for calculating the total. When the group has added the numbers, use NTO 3.11 'Calculator' to check if they were correct.

WHOLE-CLASS INTRODUCTION

3-DIGIT WITH 3-DIGIT ADDITION

You will need: BLM 37 '3-Digit Number Cards', NTO 3.7 'Modelling with MAB'

Select two cards from BLM 37 '3-Digit Number Cards' and explain to students that we want to add the numbers on the cards together, e.g. 357 + 275. Have a student model how to record the problem vertically. Model the numbers using NTO 3.7 'Modelling with MAB'. Ask, 'What do we need to do to add these numbers?' Discuss and model the process of adding the ones, tens and hundreds. Ask, 'What should we do now that there are 12 ones? Establish that 10 ones can be regrouped for 1 ten. Ask, 'What should we do now that there are 13 tens?' Model regrouping 10 tens for 1 hundred. Ask, 'What is the answer?' Repeat using different number cards. Ask, 'How would we record and solve an addition problem using three number cards?'

INDEPENDENT TASKS

Note: Choose from Tasks 1, 2 or 3.

You will need: BLM 38 'Letter Codes', deck of playing cards (picture cards, jokers and 10s removed), NTO 3.7 'Modelling with MAB', Student Book p. 60 'Addition Problems'

TASK 1: SPELLING ADDITION

Give students BLM 38 'Letter Codes'. Have them spell two- or three-letter words, such as 'up', 'cat' or 'pin'. Students use the letter codes to calculate the value of each word by forming addition problems with the letters, e.g. for 'up', students would record and solve 398 + 47 = 445. Have students try to find the two- or three-letter words with the greatest value. To challenge students further, have them calculate the value of four-, five- or six-letter words.

TASK 2: INTERACTIVE TASK 

Give pairs of students a deck of playing cards (picture cards, jokers and 10s removed). Have them turn over six cards and form two 3-digit numbers, e.g. if 6, 7, 8, 2, 5 and 1 were turned over, the numbers 678 and 251 could be formed. Students form an addition problem with the two numbers, e.g. 678 + 251. Have them use NTO 3.7 'Modelling with MAB' to model the numbers in the problem, add them together and regroup if necessary.

TASK 3: STUDENT BOOK p. 60 *'Addition Problems'*

TEACHING GROUP

You will need: BLM 37 '3-Digit Number Cards', BLM 8 'Place-Value Mat', dice, MAB, BLM 39 'Jungle Map'

REGROUPING GAME

- For students who require support, have them work in pairs and give them cards made from BLM 37 '3-Digit Number Cards', BLM 8 'Place-Value Mat', a dice and MAB. Students turn over a 3-digit number card and model this number on the place-value mat using MAB. Have students take turns rolling the dice and using the number rolled to add hundreds, tens or ones to the number on the place-value mat.

Before rolling the dice, students need to select hundreds, tens or ones, e.g. if they select tens and they roll a 6, they would add 6 tens to the number on the place-value mat. If there are then 10 or more ones or tens, the student needs to regroup and add it to the next column. The winner is the student who regroups to make 1000 exactly. If the number the student rolls brings the number past 1000, they miss their turn.

SHORTEST DISTANCE

- For students who require a challenge, present them with BLM 39 'Jungle Map'. Have them work out the shortest distance between different places on the map by forming addition problems with the distances, e.g. if travelling from Bone Cave to Treetop Mountain past the Big Boulder, the addition problem would be 107 m + 261 m = 368 m. Ask questions such as, 'What is the shortest distance between Rocky Cliff and Ghost Tree? What is the quickest way to get from Sticky Mud Swamp to Stepping Stone River?'

REFLECTION

Select from the following to suit your class and their learning outcomes:

- Have students share their work and ask, 'How do you know if an addition problem needs regrouping?' Ask, 'How is addition with 3-digit numbers similar to addition with 2-digit numbers?'
- Have students reflect on the spelling addition in Independent Task, Task 1. Ask, 'What strategies did you use to find words with a high value?'

LESSON PLAN 2

TUNING IN

WRITE A PROBLEM

You will need: large sticky notes

Write 56 – 29 on the board. Give each student a large sticky note and have them write a worded problem to match the subtraction problem. Have students swap problems and solve. Students then stick the problems on the board next to the subtraction problem. Discuss what students notice about the worded problems.

WHOLE-CLASS INTRODUCTION

3-DIGIT WITH 3-DIGIT SUBTRACTION

You will need: NTO 3.7 'Modelling with MAB'

Write the problem 573 – 298 on the board. Have a student model how to record the problem vertically. Use NTO 3.7 'Modelling with MAB' to model 573. Ask, 'What do we need to take away?' Discuss the process of subtracting the ones, tens and hundreds. Ask, 'How are we going to take away 8 ones when there are only 3 ones in the number?' Model and discuss the process of renaming 1 ten for 10 ones. Ask, 'How are we going to take away 9 tens when there are now only 6 tens in the number?' Model and discuss the process of renaming 1 hundred for 10 tens. Take away the hundreds and then ask, 'What is the answer?' Repeat with other equations, e.g. 623 – 175.

INDEPENDENT TASKS

Note: Choose from Tasks 1, 2 or 3.

You will need: BLM 4 'Blank Cards', dice, BLM 37 '3-Digit Number Cards', NTO 3.7 'Modelling with MAB', Student Book p. 61 'Subtraction Problems'

TASK 1: WORD PROBLEM MEMORY

Give each student a set of blank cards made from BLM 4 'Blank Cards'. Have students roll a dice six times and use the numbers to form two 3-digit numbers, e.g. 623 and 152. Have students record and solve a subtraction problem using their two numbers on a blank card. On another card, students write a worded problem that matches their subtraction problem. Students repeat until they have used all the blank cards, then join with a partner and play 'Memory', turning the cards face down and taking turns to turn over cards to find matching subtraction and worded problems.

TASK 2: INTERACTIVE TASK

Have students work with a partner, turning over two cards made from BLM 37 '3-Digit Number Cards' and forming a subtraction problem with the two numbers, e.g. 585 – 428. Have students use NTO 3.7 'Modelling with MAB' to model the numbers, subtracting the ones, tens and hundreds and renaming if necessary.

TASK 3: STUDENT BOOK p. 61 *'Subtraction Problems'*

TEACHING GROUP

You will need: BLM 37 '3-Digit Number Cards', MAB, BLM 12 'Sports Equipment Prices'

MAKE A PROBLEM

- For students who require support, give them cards made from BLM 37 '3-Digit Number Cards'. Have them select two cards and form a subtraction problem, then solve it by modelling with MAB. Ask, 'Did your subtraction problem have renaming?' Discuss how students know they need to rename when solving a subtraction problem. Students use other cards to solve different subtraction problems. Have them record their problems in two groups: problems with renaming and problems with no renaming.

SPORTS EQUIPMENT SHOPPING

- For students who require a challenge, give them BLM 12 'Sports Equipment Prices'. Explain that they have $600 to spend. They select equipment and then calculate how much money they would have left, e.g. if a student chooses to buy a trampoline, they solve the problem $600 – $428. To challenge students further, have them find which items they could buy to have the smallest amount of change remaining.

REFLECTION

Select from the following to suit your class and their learning outcomes:

- Have students share their work and ask, 'How do you know if a subtraction problem needs renaming?' Ask, 'How is subtraction with 3-digit numbers similar to subtraction with 2-digit numbers?'
- Have students reflect on the sports equipment shopping task in the Teaching Group. Ask, 'What strategies did you use to work out how to get the smallest amount of change?'

LESSON PLAN 3

TUNING IN

ADDITION OR SUBTRACTION

Present the problem: 'There were 84 flowers in the garden. Mrs Jones picked 28 of the biggest flowers. How many were left?' Ask, 'Is this an addition problem or a subtraction problem? How do you know?' Present the problem: '46 people travelled on the train on Tuesday and 39 people travelled on the train on Wednesday. How many people travelled on the train over the two days?' Ask, 'Is this an addition or subtraction problem?' Discuss strategies students used for determining if a problem requires addition or subtraction.

WHOLE-CLASS INTRODUCTION

A PROBLEM WITH ADDITION AND SUBTRACTION

Present the problem: 'Mrs Jacobson collects buttons and has a total of 372 buttons. She packed away her buttons into three boxes. In one box there were 134 buttons. How many buttons could be in the other two boxes?' Have students discuss how they would use addition and subtraction to solve this problem. Students work with a partner to record as many possible solutions to the problem as they can, then share their responses. Ask, 'Can you see any patterns in the answers?' Discuss the strategies students used to solve the problem.

INDEPENDENT TASKS

Note: Choose from Tasks 1, 2 or 3.

You will need: BLM 40 'Worded Problems', BLM 37 '3-Digit Number Cards', *Word,* BLM 41 'Robot Parts', Student Book p. 62 'Build a Robot'

TASK 1: WORDED PROBLEMS

Have students read and sort the worded problems from BLM 40 'Worded Problems' into two categories: addition problems and subtraction problems. Students then solve the addition and subtraction problems.

TASK 2: INTERACTIVE TASK

Give students cards made from BLM 37 '3-Digit Number Cards'. Have them select two cards and form an addition and subtraction problem using the two numbers, e.g. 428 + 363 and 428 – 363. Have students use *Word* to type worded problems to match their addition and subtraction problems. Students can then swap and answer each other's problems.

TASK 3: STUDENT BOOK p. 62 *'Build a Robot'*

TEACHING GROUP

You will need: BLM 5 'Digit Cards'

DIGIT SWAP

- For students who require support, give them cards made from BLM 5 'Digit Cards'. Have them place the cards face down and then turn over six cards to form two 3-digit numbers. Students then arrange the cards to form a vertical addition problem, e.g. 258 + 325, then rearrange the digits so they form the addition problem with the largest possible answer. The student whose answer is the largest receives a point. Students then rearrange the digits to form the subtraction problem with the smallest possible answer. The student whose answer is the smallest receives a point.

SOLVE THE PROBLEM

- For students who require a challenge, present the following problem: 'Crystal had $65 and was going to the movies. She wanted to know if she could afford to buy a large popcorn ($16) instead of a small one ($9). Crystal also had to pay $16 for her ticket. Her friend Kaitlin left her purse at home, so Crystal bought a ticket for her as well. Crystal spent $8 on a soft drink for herself and $6 on an ice-cream for Kaitlin. Crystal also needed to save $4 to catch the bus home. Did Crystal have enough money for the large popcorn?' Have students solve the problem. Ask, 'How did you use addition and subtraction to help you solve the problem?'.

REFLECTION

Select from the following to suit your class and their learning outcomes:

- Present students with the following problems: 364 + 275 and 838 – 248. Ask, 'Will these problems require regrouping or renaming?' Discuss students' strategies for determining the answer.
- Have students reflect on solving the worded problems. Ask, 'How did you know if a problem required addition or subtraction?'

Home Tasks

Select from the possible Home Tasks:

- Have students look around the home to find things that have 3-digit numbers, e.g. in books, in magazines or on packages. Have them add and subtract these numbers.
- Have students write worded addition and subtraction problems to represent their personal experiences. For example: 'I read 35 pages of my book and then I read 15 more pages. Altogether I read 50 pages.'

Assessment

- Have students complete **Student Assessment p. 63**.
- Review with students **Assessment Task Card 3.15**.

During the three lessons:

- Observe which students are able to solve 3-digit with 3-digit addition and subtraction problems with renaming or regrouping. Identify which students are able to apply their knowledge of addition and subtraction when solving problems.
- Note who completed the scaffolding tasks or the more challenging activities of the Teaching Groups.
- Review Student Book pages and make notes of areas of difficulty.

Recommendations for Future Learning

Specific to Student Assessment p. 63; if the student is experiencing some difficulty:

Q 1 Review the process of adding 3-digit with 3-digit numbers, including adding the ones, tens and hundreds. Discuss how 10 ones can be renamed as 1 ten and 10 tens can be renamed as 1 hundred. Have the student use NTO 3.7 'Modelling with MAB' to model 3-digit addition problems.

Q 2 Review the process of subtracting 3-digit with 3-digit numbers, including subtracting the ones, tens and hundreds. Discuss how 1 ten can be renamed as 10 ones and 1 hundred can be renamed as 10 tens. Have the student use NTO 3.7 'Modelling with MAB' to model 3-digit with 3-digit subtraction problems.

Q 3–4 Discuss strategies for determining if a problem requires addition or subtraction. Have the student use MAB to model the addition and subtraction problems.

Q 5 Revise how some problems require addition and subtraction.

If the student has not achieved the recommended skills for this unit:

1. See **Assessment Task Card 3.15** for specific recommendations.
2. Have the student solve addition and subtraction problems with 2-digit numbers before moving to 3-digit numbers.
3. Have the student solve addition and subtraction problems without renaming, then move to renaming.
4. Review *Nelson Maths: Australian Curriculum NSW Year 2* Unit 16.

If the student has achieved the recommended skills and these skills are firmly established, consider:

1. Having the student complete *Building Mental Strategies Skill Book Year 4*, pp. 28–29 and 40–41, to reinforce mental strategies with addition and subtraction.
2. Moving forward to *Nelson Maths: Australian Curriculum NSW Year 4* Unit 13.
3. Extending the student in any of the listed activities by using larger numbers or numbers involving decimals.

Unit 16 Time

Measurement and Geometry

Time MA2-13MG reads and records time in one-minute intervals and converts between hours, minutes and seconds

analogue, clock, digital, half past, hours, minutes, minutes past, o'clock

LESSON PLAN 1

TUNING IN

HOW LONG IS ONE MINUTE?

You will need: stopwatch or NTO 3.12 'Stopwatch'

Make sure no clocks are visible and have all students stand. Say, 'I'm going to time one minute. When you think one minute has passed, sit down'. On the minute, take note of students who were the most accurate, but keep timing. Stop the clock at about 1 min 30 sec. Ask the students who were the closest to the minute mark to share any strategies with the class. Repeat the activity.

WHOLE-CLASS INTRODUCTION

WHAT IS ONE MINUTE?

You will need: NTO 3.18 'Clocks'

Display NTO 3.18 'Clocks' and select the time as half-past 1. Briefly discuss the analogue clock. Point to the hour hand and ask, 'What does this hand tell us?' Repeat for the minute hand. Select the 'seconds on' option. Discuss the second hand, explaining that not all clocks have this hand. Ask, 'Which hand moves the fastest?' Discuss. Select the 'add time' option and 'add 1 minute'. Watch the second hand move around the clock for one minute, draw students' attention to the minute hand as it moves once after the second hand has completed its circuit. Ask, 'How many times did the second hand move to complete a full circuit?' Discuss, then add another minute and have students count the movements. Emphasise the idea that there are 60 seconds in one minute.

INDEPENDENT TASKS

Note: Choose from Tasks 1, 2 or 3.

You will need: *Word,* BLM 42 'Typing to the Minute', Student Book p. 64 'In One Minute ...'

TASK 1: GETTING READY IN MINUTES

Have students make a detailed list of what they do from the time they wake up until they get to school, e.g. get out of bed, walk to kitchen, make breakfast. Next to each, have them write whether the activity takes 'less than one minute', 'about one minute' or 'more than one minute'.

TASK 2: INTERACTIVE TASK 

Have students work in pairs using *Word* to complete BLM 42 'Typing to the Minute'. Students predict how many times they can type words and sentences in one or two minutes, then take turns to time or type and see how accurate their predictions were.

TASK 3: STUDENT BOOK p. 64 *'In One Minute ...'*

TEACHING GROUP

You will need: NTO 3.18 'Clocks', stopwatches

UNDERSTANDING ONE MINUTE

- For students who require support, spend more time looking at the movement of the second and minute hands on NTO 3.18 'Clocks' to reinforce that there are 60 seconds in one minute. Then, give each student a stopwatch. Have them stop the stopwatch as close to one minute as they can. Compare and discuss results and strategies. Repeat, looking for improvement. Ask, 'How many times can you clap in one minute?' Have students make a prediction and then work with a partner and a stopwatch to see how many times they can. Repeat with other activities, e.g. star jumps, saying the alphabet and writing their name.

EXTENDING TO MINUTES

- For students who require a challenge, have them extend their understanding by looking at the class timetable and selecting ten events, then classifying them into the following categories: 'less than 10 min, < 20 min, < 30 min, < 60 min'. As a further extension, have them work out the number of seconds in each of the minutes listed, e.g. 10 min = 600 seconds.

REFLECTION

Select from the following to suit your class and their learning outcomes:

- As a class, play 'How Long Is One Minute?' from Tuning In again. As a challenge, extend the time to two minutes.
- Ask, 'Where do we use seconds and minutes in everyday life?'

LESSON PLAN 2

TUNING IN

PEGS ON A LINE TO SHOW TIME

You will need: BLM 43 'Time Cards', magnets

Draw a line on the whiteboard, about 150 cm long. Use a magnet to attach 11:00 from BLM 43 'Time Cards' to one end of the line and 12:00 to the other. Ask, 'What are these times?' Students should say the times as o'clock. Ask, 'What are some times that come between these times?' Discuss. Hold up the 11.30 card and ask, 'Where would this go?' Have a student attach it to the line. Repeat with 11.15, 11.45 and 11.05, discussing where they would sit and why. Use language such as 'minutes past', 'before' and 'after' when describing and comparing times. This timeline is to be used in the Whole-Class Introduction.

WHOLE-CLASS INTRODUCTION

READING AND UNDERSTANDING MINUTES

You will need: NTO 3.18 'Clocks', BLM 43 'Time Cards'

Display NTO 3.18 'Clocks' and move the clock hands to show 11 o'clock. Ask, 'What time is this? How do you know it is something o'clock?' Students should recognise that o'clock means the minute hand is on the 12. Look at the timeline from Tuning In and ask, 'What does 12 o'clock look like?' Have a student volunteer to show 12:00 on the NTO. Ask, 'How many minutes are there between 11 and 12 o'clock, or in one hour?' Discuss, emphasising 60 minutes. Ask, 'How many minutes are there between each number on the clock face?' Count the pointers between 12 and 1 to show that there are five. Count by 5s around the clock face to show 60 minutes in total. Click on 'show minute values' to display minutes. Hand out the remaining cards from BLM 43 'Time Cards' and have students show times on the clock, then add the cards to the line on the board.

INDEPENDENT TASKS

Note: Choose from Tasks 1, 2 or 3.

You will need: BLM 44 'Match the Time', workbook/sheet of paper, LO: *L9648 'Time tools: 12-hour to the minute: practice time'*, BLM 43 'Time Cards' (optional), Student Book p. 65 'Minutes on a Clock'

TASK 1: MATCH THE TIME

Give students BLM 44 'Match the Time' and have them follow the instructions to match digital and analogue times between 3:00 and 4:00 to the minute. Students paste the work into their workbooks or onto a sheet of paper, then write the times in words beside each clock.

TASK 2: INTERACTIVE TASK 

Have students work independently on computers, using LO: *L9648 'Time tools: 12-hour to the minute: practice time'* to manipulate the time on a clock, adding or subtracting time and watching the hands move. Students could model times from BLM 43 'Time Cards' or select their own times.

TASK 3: STUDENT BOOK p. 65 ***'Minutes on a Clock'***

TEACHING GROUP

You will need: NTO 3.18 'Clocks' or small learning-aid clocks, BLM 43 'Time Cards', BLM 45 'Blank Clock Faces'

MINUTES CHALLENGE

- For students who require support, use NTO 3.18 'Clocks' or small learning aid clocks on which students can manipulate the time. Emphasise that there are 60 minutes in an hour and count by 5s around the clock. Have students work in pairs and take turns to model the times on BLM 43 'Time Cards'. For each time they get right the first time they get 5 points; for two tries, 2 points; and for three tries, 1 point.

ADDING TIME

- For students who require a challenge, have them work in pairs with a series of cards from BLM 43 'Time Cards' and clock faces from BLM 45 'Blank Clock Faces'. Students create 'adding time' challenges for each other by selecting a time card and choosing a number of minutes to add to the time. The new time is then drawn on the clock face.

REFLECTION

Select from the following to suit your class and their learning outcomes:

- Using NTO 3.18 'Clocks', point to any number (1–12) and have students say the number of minutes this represents.
- Make a set of cards from BLM 44 'Match the Time' and give one to each student. Without talking, students find their matching time (digital/analogue), then line up in order.

LESSON PLAN 3

TUNING IN

TIME IN OUR LIVES

Have students brainstorm where they use or hear people talk about minutes, e.g. 'I'll just be a minute', 'Wait a minute'. Do the same for hours, e.g. basketball practice is one hour, music lessons are half an hour. Ask, 'How are these two lists similar or different? Hours and minutes are both measures of time – are they used in the same way?' If there is time, discuss seconds and where we use them, e.g. for timing races. Ask, 'Why don't we use minutes?'

WHOLE-CLASS INTRODUCTION

THE HOUR HAND

You will need: NTO 3.18 'Clocks'

Display NTO 3.18 'Clocks' and ask, 'If it is 6.30, where would the hour hand be pointing?' Discuss, explaining that the hour hand moves slowly as the minutes pass. Show 6.30 on the IWB, pointing out that the hour hand is exactly half way between the 6 and 7. Ask, 'Where do you think the hour hand would be at 6.45?' Discuss, and show on the IWB. Select other times, e.g. 3:10 and 8:55, to show the position of the hour hand. Then point to a position on the clock, and ask, 'If the hour hand was pointing here, what would the time be?' Unlike the minute hand, the hour hand can tell the time on its own – but having the minute hand makes it simpler and more accurate.

INDEPENDENT TASKS

Note: Choose from Tasks 1, 2 or 3.

You will need: BLM 46 'Making a Clock', coloured card, split pins, BLM 47 '*PowerPoint* Clocks', *PowerPoint*, Student Book p. 66 'Analogue Time'

TASK 1: MAKING A CLOCK

Give each student BLM 46 'Making a Clock' photocopied onto coloured card and a split pin. Have them make their clock then work with a partner, taking turns to make and guess times.

TASK 2: INTERACTIVE TASK 

Give each student BLM 47 '*PowerPoint* Clocks' and have them work independently on *PowerPoint* to complete this activity.

TASK 3: STUDENT BOOK p. 66 *'Analogue Time'*

TEACHING GROUP

You will need: BLM 45 'Blank Clock Faces', access to the internet or encyclopaedias

TIME IN MY LIFE

- For students who require support, ask them to make a list of ten things they usually do in a day, e.g. eat breakfast, drive to school. Next to each item, have them write down in digital format an approximate time when they start this activity. Then give students clock faces from BLM 45 'Blank Clock Faces' and have them fill in the time, reminding them how to use the hour and minute hands.

SUNDIALS

- For students who require a challenge, have them research ancient methods of telling the time, e.g. sundials. After finding out how these work, have students liken to the hour hand on a clock – being able to tell the time with a single line.

REFLECTION

Select from the following to suit your class and their learning outcomes:

- Invite students who completed Independent Tasks, Task 2, to show their *PowerPoint* presentations to the class. They may like to hide the digital time and have the class guess the time shown.
- Ask, 'If we can tell time with the hour hand alone, why do we need the minute hand?'

Home Tasks

Select from the possible Home Tasks:

- Have students make a list of all the time-telling devices in their home, where they are located and whether they are analogue or digital, e.g. digital clock on oven in kitchen, analogue clock on wall in family room.
- Have students make a list of all of the activities they do on one day of the weekend, then put them into categories of their duration: 'less than 1 hour', 'about 1 hour' and 'more than 1 hour'.

Assessment

- Have students complete **Student Assessment p. 67**.
- Review with students **Assessment Task Card 3.16**.

During the three lessons:

- Make note of how students go about working with time devices – are they confident in using the stopwatch and moving/drawing the hands on the clock?
- Collect students' *PowerPoint* presentations from Lesson Plan 3, Independent Tasks, Task 2, to show the use of ICT to enhance leaning.
- Review Student Book pages and make note of areas of difficulty.

Recommendations for Future Learning

Specific to Student Assessment p. 67; if the student is experiencing some difficulty:

Q 1 Revise how many seconds there are in a minute and how many minutes in an hour. Use an analogue clock to demonstrate this, having the student count the movement of the second hand if required.

Q 2 Revisit the 'What Is One Minute?' activity from Lesson Plan 1, Whole-Class Introduction. Discuss times that would be 'about one minute' – between 50 and 70 seconds. Times could also be placed on a number line to show their distance from one minute.

Q 3–4 Revise the 'Reading and Understanding Minutes' activity from Lesson Plan 2, Whole-Class Introduction. Have the student use a learning-aid clock to manipulate the time. Revise the strategy of counting by 5s to find minutes – use NTO 3.18 'Clocks' with the 'show minute values' option selected. Have the student say times aloud and look carefully at the clocks in deciding earliest and latest times.

Q 4 Encourage the student to think about the movement of the hour hand. Revisit activities in Lesson Plan 3.

Q 5 Have the student apply logical reasoning to think about how they use time in their own lives.

If the student has not achieved the recommended skills for this unit:

1. See **Assessment Task Card 3.16** for specific recommendations.
2. Have the student work with hands-on materials, such as learning-aid clocks where they can move the hands themselves or NTO 3.18 'Clocks'.
3. Have the student complete matching activities between digital and analogue clocks, similar to BLM 44 'Match the Time'. These cards could be used in games such as 'Snap' to reinforce understanding.
4. Review *Nelson Maths: Australian Curriculum NSW Year 2* Units 14 and 21.

If the student has achieved the recommended skills and these skills are firmly established, consider:

1. Having the student convert times (seconds to minutes, minutes to hours and 12-hour to 24-hour format).
2. Moving forward to *Nelson Maths: Australian Curriculum NSW Year 4* Units 25 and 26.
3. Having the student read times after half past as 'minutes to' the hour.

Unit 17 Angles

Measurement and Geometry
Angles MA2-16MG identifies, describes, compares and classifies angles

 angle, right angle, curved, diagonal, horizontal, line, parallel lines, quarter turn, rotate, straight, vertical

LESSON PLAN 1

TUNING IN

WHAT TYPES OF LINES?

As a group, brainstorm the types of lines that students are familiar with. Record suggestions on the board and invite students to draw examples of the different types of lines. Ensure students discuss straight, curved, diagonal, horizontal and vertical lines.

WHOLE-CLASS INTRODUCTION

BODY LINES

Have students discuss how they could use their body to model the different types of lines discussed in Tuning In, e.g. lying straight on the floor would represent horizontal and standing up straight would represent vertical. Tell students that as you say the name of a type of line, they need to make that line with their body. Choose from the following lines: curved, diagonal, horizontal and vertical.

INDEPENDENT TASKS

Note: Choose from Tasks 1, 2 or 3.

You will need: BLM 48 'Label the Lines', *Word,* Student Book p. 68 'Drawing Lines'

TASK 1: LABEL THE LINES

Give each student BLM 48 'Label the Lines'. Have them label the lines as curved, diagonal, horizontal or vertical, then draw examples of objects that have curved, diagonal, horizontal or vertical lines.

TASK 2: INTERACTIVE TASK 

Have students work with a partner drawing lines using *Word.* One student draws a curved, diagonal, horizontal or vertical line, then their partner writes the label for that line. Students then swap roles.

TASK 3: STUDENT BOOK p. 68 ***'Drawing Lines'***

TEACHING GROUP

You will need: sticky notes, rulers

LOOKING FOR LINES

- For students who require support, make a list of the lines that have been introduced: straight, curved, diagonal, horizontal and vertical. Have students draw examples of each type of line. Give them sticky notes and have them find examples of the lines around the room and label them using the sticky notes.

PARALLEL LINES

- For students who require a challenge, introduce them to parallel lines. Explain that parallel lines are two straight lines that are always the same distance apart. Have students use a ruler to practise drawing parallel lines. Ask, 'Would parallel lines ever join up? Why not?' Have students record a list of the parallel lines they can find in the classroom.

REFLECTION

Select from the following to suit your class and their learning outcomes:

- Have students recall the different types of lines. Ask, 'Which type of line is the most common type of line in our classroom?'
- Have students reflect on the Independent Tasks and the Teaching Group activities. Encourage them to share what they have learned about lines.

LESSON PLAN 2

TUNING IN

WHAT IS AN ANGLE?

You will need: two rulers

An angle is an opening or turn. One way to help students to understand this is to use two rulers as the arms of an angle. Place one on top of the other against the board. Holding one end together, pivot the upper ruler upwards and identify the turn as an angle. Trace the angle made by the rulers on the board. When the rulers are moved away the students can see the angle and how it was formed.

WHOLE-CLASS INTRODUCTION

RIGHT ANGLE SEARCH

Repeat the Tuning In activity and stop the turn at a right angle. Identify this as a special angle called a right angle. Ask students to look around them and to find as many examples of right angles as they can. Tell them they are likely to find hundreds of examples.

INDEPENDENT TASKS

Note: Choose from Tasks 1, 2 or 3.

You will need: paper or card, Student Book p. 69 'Spot the Right Angles', NTO 3.19 'Rotating Lines'

TASK 1: HUNT FOR ANGLES

Have students make a 'right-angle tester' from the corner of a rectangular piece of paper or card. Ask them to hunt for examples of angles around the classroom that are less than, more than or exactly right angles. Have them draw examples of each type of angle and record where they found them.

TASK 2: INTERACTIVE TASK 

Have students use NTO 3.19 'Rotating Lines' to create and identify angles that are less than, more than or exactly right angles.

TASK 3: STUDENT BOOK p. 69 *'Spot the Right Angles'*

TEACHING GROUP

You will need: a right-angle tester for each student, angles drawn on a large sheet of paper

LOOK FOR ANGLES

- For students who require support, prepare a sheet of paper with large acute, obtuse and right angles drawn on it. Help the student to place their right-angle tester accurately on the vertex of the angle to test whether it is less than, more than or exactly a right angle.

ANGLE CHALLENGE

- For students who require a challenge, ask them to use a ruler and a pencil to draw an angle that is as close as possible to a right angle. Next have them use the right-angle tester on it and describe the strategy they used to draw the right angle.

REFLECTION

Select from the following to suit your class and their learning outcomes:

- Have students reflect on their angle hunt in Independent Tasks, Task 1. Ask, 'Where did you find the most angles? What was the most common angle that you found?'
- Have students share what they have learned about angles. Ask, 'What is an angle? How is an angle formed?'

TUNING IN

ROTATE THE LINES

You will need: two strips of paper joined at the ends with a split pin to enable angles to be formed

Present the joined strips of paper as a straight line to students. Have them identify the angle that they see. Rotate one of the strips of paper and have students discuss what is happening. Discuss how angles are formed as lines rotate. As you rotate the strips of paper, emphasise how angles of different sizes are formed. Then present students with an angle using the two strips of paper. Ask, 'What can you see around the room that matches this angle?'

WHOLE-CLASS INTRODUCTION

QUARTER TURN

You will need: NTO 3.19 'Rotating Lines'

Use NTO 3.19 'Rotating Lines' to introduce the concept of angles that are equal to a quarter turn. Rotate one of the lines a quarter turn. Have students describe the angle that they see. Ask, 'How could we rotate these lines to make the same angle?' Discuss students' ideas. Select a student to demonstrate how the two lines can be turned to form a quarter turn in a different position. Ask, 'Where have you seen an angle like this before? Can you see any angles like this around the room?' Discuss how the rotating lines are similar to the hands of a clock.

INDEPENDENT TASKS

Note: Choose from Tasks 1, 2 or 3.

You will need: strips of paper, split pins, NTO 3.19 'Rotating Lines', Student Book p. 70 'Angle Time'

TASK 1: SEARCH FOR QUARTER TURN ANGLES

Have students make their own rotating angles by joining the ends of two strips of paper with a split pin. Have students rotate one of the strips of paper to form an angle that is equal to a quarter turn. Students search around the room for angles that are equal to a quarter turn. Encourage them to place their strips of paper against objects around the room to compare the size of the angles. Have students draw the angles they find and record where they found them.

TASK 2: INTERACTIVE TASK 

Have students work with a partner using NTO 3.19 'Rotating Lines'. One student forms an angle by rotating the two lines. Their partner then tries to find an example of that angle in the room. Students then swap roles.

TASK 3: STUDENT BOOK p. 70 *'Angle Time'*

TEACHING GROUP

You will need: geoboards, elastic bands, clocks with movable hands

SHAPE ANGLES

- For students who require support, have them make shapes using geoboards and elastic bands. Have them identify the angles in their shapes, then move the elastic bands to change the angles. Discuss how the angles change as the lines in the shape rotate or move. Have students identify if there are any angles in their shapes that are equal to a quarter turn.

MORE CLOCKS AND ANGLES

- For students who require a challenge, show them a clock with movable hands. Have students explore how different angles are formed as the hands of the clock move. Discuss how the angles can be different sizes. Have students move the hands of the clock to make different times. Have them record and compare the angles that they form. Ask, 'What time forms the biggest/smallest angle?'

REFLECTION

Select from the following to suit your class and their learning outcomes:

- Have students reflect on the Independent Tasks and Teaching Group activities. Have them share what they have learned about angles.
- Have students look at the classroom door. Open and close it and have students describe the angle they see. Ask, 'How is the angle changing as the door opens and closes?'

Home Tasks

Select from the possible Home Tasks:

- Have students search for and make a list of angles they can see in each room of the house.
- Have students look at an analogue clock and describe the angle formed by the two hands. Have them identify the times on the clock when the hands make a right angle.

Assessment

- Have students complete **Student Assessment p. 71**.
- Review with students **Assessment Task Card 3.17**.

During the three lessons:

- Observe which students are able to identify and draw different angles.
- Make note of students who completed the scaffolding tasks or the more challenging activities of the Teaching Groups.
- Review Student Book pages and make notes of areas of difficulty.

Recommendations for Future Learning

Specific to Student Assessment p. 71; if the student is experiencing some difficulty:

Q 1 Revise the different types of lines: curved, vertical, horizontal and diagonal. Have the student explain how the lines are similar and different, and practise identifying and drawing each type of line.

Q 2–3 Revise how an angle is formed. Have the student draw two straight lines that meet and identify the angle formed as less than, more than or exactly a right angle. Have the student identify angles around the room.

Q 4 Revise the concept of an angle being the rotation of a line. Have the student attach two strips of paper with a split pin and then rotate the lines to recognise how different angles are formed. Have the student rotate one strip of paper a quarter turn and describe the angle as a right angle. Have the student find other angles that are equal to a right angle.

If the student has not achieved the recommended skills for this unit:

1. See **Assessment Task Card 3.17** for specific recommendations.
2. Have the student use materials such as straws and masking tape to make models of different types of angles.
3. Review *Nelson Maths: Australian Curriculum NSW Year 2* Unit 8.

If the student has achieved the recommended skills and these skills are firmly established, consider:

1. Introducing the concept of acute and obtuse angles and having them compare and order angles according to size.
2. Moving forward to *Nelson Maths: Australian Curriculum NSW Year 4* Unit 22.

Unit 18 Mental Strategies for Multiplication

Number and Algebra
Multiplication and division MA2-6NA uses mental and informal written strategies for multiplication and division

double, fact, groups of, lots of, multiply, ones, place-value, repeated addition, skip count, strategy, tens, times table

LESSON PLAN 1

TUNING IN

WHAT IS MULTIPLICATION?

You will need: counters

Write the multiplication symbol on the board. Ask, 'What does this sign mean?' Record responses, e.g. groups of, rows of, piles of, arrays. Encourage students to use counters to demonstrate multiplication problems.

WHOLE-CLASS INTRODUCTION

MULTIPLICATION AND SKIP COUNTING

You will need: NTO 3.20 'Counters'

Have students count by 2s up to 20 and record the sequence on the board. Then use NTO 3.20 'Counters' to make a group of two counters. Discuss how there is one group of 2 and model how it can be recorded as $1 \times 2 = 2$. Add another group of two counters and discuss how there are two groups of 2 and record as $2 \times 2 = 4$. Continue adding groups of 2 and recording the equations. Refer back to the counting by 2s sequence and discuss what students notice. Repeat for the 10 and 5 times tables, recording the skip-counting pattern, modelling the groups with counters and recording the multiplication equations. Ask, 'How could you use the skip-counting pattern as a strategy for solving multiplication problems?'

INDEPENDENT TASKS

Note: Choose from Tasks 1, 2 or 3.

You will need: 10-sided dice, playing cards (2s, 5s and 10s only), NTO 3.9 '100 Chart', counters, Student Book p. 72 'At the Supermarket'

TASK 1: SKIP COUNT TO SOLVE

Have students roll a 10-sided dice and turn over a card from a pile of playing cards (2s, 5s and 10s only). Students record a multiplication problem using the two numbers, e.g. 7×5. Have them solve the problem using the skip-counting sequence, e.g. for 7×5, students would skip count by 5 seven times to 35. Encourage students to draw pictures or use counters to model the skip-counting pattern.

TASK 2: INTERACTIVE TASK

Present students with NTO 3.9 '100 Chart' (show all numbers). Have them highlight the numbers that are part of the counting-by-2s sequence up to 20. Have them model the highlighted numbers by making groups of two counters. Students then write the multiplication fact that would match, e.g. $1 \times 2 = 2$, $2 \times 2 = 4$. Students repeat the task for 5s and 10s sequences.

TASK 3: STUDENT BOOK p. 72 *'At the Supermarket'*

TEACHING GROUP

You will need: counters, blank pieces of card

REPEATED ADDITION

- For students who require support, present them with the problem 7×2. Have them use counters to model the problem by making seven groups of 2. Have students demonstrate how they can skip count to find the total amount. Ask, 'How could we use addition to solve this problem?' Discuss how the problem $2 + 2 + 2 + 2 + 2 + 2 + 2 = 14$ matches the multiplication problem. Present the following problems: 6×1, 9×10, 8×2, 5×1, 7×10 and 3×2. Have students solve these problems by modelling with counters, skip counting and using repeated addition.

SKIP COUNT STOP

- For students who require a challenge, have them work with a partner. One student writes a multiplication fact from the 2, 5 or 10 times tables on a card without their partner seeing. They then ask their partner to skip count by 2s, 5s or 10s, depending on the multiplication fact that they wrote. When their partner says the number that is the answer to the multiplication fact on the card, the student says, 'Stop'. Their partner then works out the multiplication fact. Students swap roles.

REFLECTION

Select from the following to suit your class and their learning outcomes:

- Have students reflect on the Independent Tasks. Ask, 'How did skip counting help you solve the multiplication problems?'
- Present the problems: 6 × 5, 7 × 2, 4 × 10. Have students explain the strategy they would use to solve them.

LESSON PLAN 2

TUNING IN

DOUBLE CALL OUT

You will need: NTO 3.10 'Card Flip'

Revise the concept of doubles. Present students with NTO 3.10 'Card Flip', selecting 0–20 cards. Ask three students to stand up. Explain that as a number is shown, they need to double it and call out the answer. The first to say the correct answer wins a point. The winner is the first to get 5 points. Repeat with different students.

WHOLE-CLASS INTRODUCTION

MULTIPLYING BY 2 AND DOUBLES

You will need: counters, NTO 3.17 'Ten-Sided Dice'

Have students share what they know about the 2 times table. Discuss how it means 'groups of 2'. Present the problem: 4 × 2. Ask, 'What can you tell me about this problem?' Have a student use counters to model four groups of 2 and record the answer 8 on the board. As a group, count the four groups and emphasise how there are two counters in each group. Ask, 'How could we use our knowledge of doubles to help us multiply numbers by 2?' Write 'double 4' on the board. Have students solve the doubles problem and record the answer. Ask, 'What do you notice about the answers to 4 × 2 and double 4? Why would they have the same answer?' Have students work with a partner. Present a number using NTO 3.17 'Ten-Sided Dice'. Have one student multiply the number by 2 while their partner doubles the number. They compare their answers. Repeat with different numbers.

INDEPENDENT TASKS

Note: Choose from Tasks 1, 2 or 3.

You will need: BLM 4 'Blank Cards', dice, BLM 5 'Digit Cards', *Word* and *ClipArt*, Student Book p. 73 'Doubles and Multiplying by 2'

TASK 1: MATCHING CARDS

Give students cards made from BLM 4 'Blank Cards' and a dice. Have students take three blank cards and roll the dice. On one card they record that number in a 2 times table, on another card they record that number as a doubles problem and on the third card they record the answer. For example, if the student rolled a 5, they would record '5 × 2', 'double 5' and '10' on the cards. Students continue by rolling the dice again and recording other problems and answers. When they have used all their cards, have students shuffle and then match their cards. They could then join their cards with a partner's and play a game of 'Memory' by placing the cards face down and turning over three cards at a time to match a 2 times table, a doubles problem and an answer.

TASK 2: INTERACTIVE TASK 

Give students cards made from BLM 5 'Digit Cards'. Have them take a card and using *Word,* type a 2 times table and a doubles fact using the number on their card. For example, for 7, they would type '7 × 2 = 14' and 'double 7 is 14'. Students could use *ClipArt* pictures to represent the times table and doubles problem.

TASK 3: STUDENT BOOK p. 73 *'Doubles and Multiplying by 2'*

TEACHING GROUP

You will need: Unifix blocks

TOWERS OF TWO OR TWO TOWERS

- For students who require support, present them with the problem: 3 × 2. Have them use Unifix blocks to make towers representing the equation. Ask, 'How did you know that you needed three towers? Why did you put two blocks in each tower?' Record the problem: double 3. Have students use the blocks to model this equation. Ask, 'Why did you make two towers? Why did you put three blocks in each tower?'

Have students discuss what they notice about the two problems. Discuss how one had three towers with two blocks in each and the other had two towers with three blocks in each. Emphasise that the two problems had the same answer. Repeat with other problems, e.g. 8 × 2, 10 × 2, 2 × 2.

BIGGER NUMBERS

- For students who require a challenge, present the problem: 13 × 2. Ask, 'How could you use doubles facts to help you solve this multiplication problem?' Discuss the place value of the digits in the number 13. Establish that 13 is a ten and 3 ones. Discuss how we can double each of these and then add them together to work out double 13, e.g. double 10 is 20 and double 3 is 6, so double 13 would be 26. Present students with the following numbers: 37, 13, 24, 46, 26, 31 and 52. Have students multiple these numbers by 2.

REFLECTION

Select from the following to suit your class and their learning outcomes:

- Have students share their work. Ask, 'How can you use doubles facts to help you multiply numbers by 2?'
- Have students answer 2 times tables facts as quickly as possible. Ask, 'What strategy did you use?'

LESSON PLAN 3

TUNING IN

GROUPS OF 1, GROUPS OF 10

You will need: counters

Have students work with a partner. Explain that they are going to make groups of 1 and groups of 10. Present the problem: 4 × 1. Have students model the equation using counters. Ask, 'How many counters were in each group?' Present the problem: 4 × 10. Have students model this with their counters. Ask, 'How many counters were in each group?' Discuss what students notice about the two problems. Repeat with 7 × 1 and 7 × 10.

WHOLE-CLASS INTRODUCTION

USING PLACE VALUE

You will need: NTO 3.7 'Modelling with MAB'

Write the problem 6 × 1 on the board. Ask, 'How would we model this problem using MAB?' Have a student model the problem using NTO 3.7 'Modelling with MAB' and emphasise how there are six lots of 1. Present the problem: 6 × 10. Say, 'What should we use to model lots of 10?' Have a student model the problem with 6 tens. Repeat with other problems, e.g. 3 × 1, 3 × 10. Emphasise that we use ones to multiply by 1 and tens to multiply by 10.

INDEPENDENT TASKS

Note: Choose from Tasks 1, 2 or 3.

You will need: BLM 49 'Rocket Race', 10-sided dice, NTO 3.7 'Modelling with MAB', 10-sided dice, BLM 5 'Digit Cards', Student Book p. 74 'Multiply by 1, Multiply by 10'

TASK 1: ROCKET RACE

Have students play with a partner using BLM 49 'Rocket Race'. One students rolls the 10-sided dice, then chooses to multiply the number shown by 1 or by 10. The student then colours a space in their rocket that is the answer for their multiplication problem. If there are no spaces that match, they miss their turn. Students continue the game, taking turns. The winner is the first student to colour all spaces in their rocket.

TASK 2: INTERACTIVE TASK 

Have students work with a partner, turning over a card made from BLM 5 'Digit Cards'. Students multiply that number by 1 and by 10 and model the problems using NTO 3.7 'Modelling with MAB'. They then record the two equations.

TASK 3: STUDENT BOOK p. 74 ***'Multiply by 1, Multiply by 10'***

TEACHING GROUP

You will need: BLM 31 'Counting by 10 Cards', BLM 5 'Digit Cards', MAB, NTO 3.7 'Modelling with MAB'

MODEL THE NUMBER

- For students who require support, give them cards made from BLM 31 'Counting by 10 Cards' and BLM 5 'Digit Cards'. Combine and shuffle the cards and then turn a card over. Have students use MAB to model the number, e.g. for 40, students would use 4 tens. Discuss how students know whether to model the number with tens or ones. Have students record a multiplication problem to match the MAB, e.g. 4 × 10 = 40. Continue, using different cards. Discuss how equations that were modelled with tens were written with ' × 10' and equations that were modelled with ones were written with ' × 1'.

MULTIPLYING MULTIPLES OF 10

- For students who require a challenge, present the problem: 36 × 20. Ask, 'How could we solve this problem?' Discuss students' ideas. Ask, 'Could we use the problem 36 × 10 to help us solve 36 × 20?

Use NTO 3.7 'Modelling with MAB' to demonstrate how multiplying by 20 is doubling the answer of multiplying by 10. Discuss how multiplying 36 by 2 and then by 10 will also solve the problem. Provide the following problems for students to solve: 16 × 30, 25 × 20 and 18 × 40. Discuss the strategies they used to solve the problems.

REFLECTION

Select from the following to suit your class and their learning outcomes:

- Have students share their work and ask, 'How did your knowledge of place value help you with multiplying numbers by 1 or by 10?'
- Say facts from the 1 and 10 times tables, e.g. 4 × 10 or 6 × 1. Have students answer the problems as quickly as possible. Ask, 'What strategy did you use to help you answer the multiplication problems?'

Home Tasks

Select from the possible Home Tasks:

- Have students find examples of multiplication facts around the home, e.g. 6 plates with 2 potatoes on each plate is 6 × 2 = 12.
- Have students find single-digit numbers around the home and multiply them by 2, 1 and 10. Ask them to explain the strategies they used to solve the equations.

Assessment

- Have students complete **Student Assessment p. 75**.
- Review with students **Assessment Task Card 3.18**.

During the three lessons:

- Observe which students are able to use the mental strategies to solve the multiplication facts.
- Note who completed the scaffolding tasks or the more challenging activities of the Teaching Groups.
- Review Student Book pages and make notes of areas of difficulty.

Recommendations for Future Learning

Specific to Student Assessment p. 75; if the student is experiencing some difficulty:

Q 1 Review the strategies for multiplying numbers by 2, 1 and 10. Provide opportunities for the student to use the strategies to solve multiplication facts from the 2, 1 and 10 times tables.

Q 2 Revise how doubles facts can be used to solve multiplication facts when a number is multiplied by 2. Have the student solve 2 times table facts by modelling groups of 2 and also by doubling the number. Emphasise that the answer is the same.

Q 3 Revise how an understanding of place value can help when multiplying numbers by 1 and 10. Have the student model 1 and 10 times tables facts using MAB. Emphasise that lots of 1 are made using ones and lots of 10 are made using tens.

Q 4 Revise the mental strategies for multiplication. Have the student make a poster of the strategies that they know how to use. Encourage the student to refer to the poster when solving problems.

Q 5 Revise how skip counting can be used to solve multiplication problems. Have the student count by 1s, 2s and 10s, then solve 2, 1 and 10 times tables facts and identify the counting patterns in the answers.

If the student has not achieved the recommended skills for this unit:

1. See **Assessment Task Card 3.18** for specific recommendations.
2. Have the student use materials to model the multiplication problems. Encourage them to solve them by skip counting, using repeated addition and the mental strategies.
3. Review *Nelson Maths: Australian Curriculum NSW Year 2* Unit 23.
4. Have the student complete *Building Mental Strategies Skill Book Year 3*, pp. 58–62, to reinforce mental strategies for multiplication.

If the student has achieved the recommended skills and these skills are firmly established, consider:

1. Having the student complete *Building Mental Strategies Skill Book Year 4*, pp. 48–50, to further explore mental strategies for multiplication.
2. Moving forward to *Nelson Maths: Australian Curriculum NSW Year 4* Unit 10.
3. Extending the student by multiplying numbers by 5 and 3.

Unit 19 More About Mental Strategies for Multiplication

Number and Algebra
Multiplication and division MA2-6NA uses mental and informal written strategies for multiplication and division

addition, double, groups of, halve, MAB, multiply, number sequence, skip counting, strategy, times tables

LESSON PLAN 1

TUNING IN

HALVING

Have students walk around the room. Say an even number, e.g. 10, and have them stand in groups of that number. Then have half the students in each group sit down. The quickest group wins that round. Discuss strategies students used for calculating half of the group.

WHOLE-CLASS INTRODUCTION

MULTIPLY BY 10 AND HALVE

As a group, record the 10 times table, with answers, on the board. Next to this, write the 5 times table, without answers. Students solve the first two facts, then use skip counting to solve the rest. Have students look closely at both tables. Ask, 'What do you notice about the answers of the 10 times table and the answers of the 5 times table?' Discuss how the 5 times table answers are half of the 10 times table answers. Explain how this means that when multiplying a number by 5, we can multiply it by 10 and then halve it to quickly work out the answer. Present the equation: 4 × 5. Model solving it by multiplying 4 by 10 to make 40 and then halving 40 to make 20 (i.e. 4 × 10 = 40 and half of 40 is 20). Present equations from the 5 times table and have students solve them using this strategy.

INDEPENDENT TASKS

Note: Choose from Tasks 1, 2 or 3.

You will need: 10-sided dice, calculator, NTO 3.17 'Ten-Sided Dice', Student Book p. 76 'Multiply by 5'

TASK 1: HIGHEST TOTAL

Students play with a partner and take turns to roll a 10-sided dice, multiply the number shown by 5 and record their answer. Have students continue until they have each rolled the dice 10 times. Each student then finds the total of their answers, using a calculator. The winner is the student with the highest total. Have students play the game again.

TASK 2: INTERACTIVE TASK 

Present students with NTO 3.17 'Ten-Sided Dice' and have them generate a number using the dice, then multiply the number by 5 using the mental strategy 'multiply by 10 and halve'. Have students compare their answer with a partner's to ensure they are the same.

TASK 3: STUDENT BOOK p. 76 *'Multiply by 5'*

TEACHING GROUP

You will need: BLM 5 'Digit Cards', Unifix blocks, NTO 3.7 'Modelling with MAB'

HALVING TOWERS

- For students who require support, have them work in pairs with cards made from BLM 5 'Digit Cards' and Unifix blocks. Place the cards face down and have students turn over a card, then multiply the number shown by 5. Have one student model the equation making groups of 5, e.g. making towers of five Unifix blocks. Their partner solves the equation by multiplying the number by 10, e.g. making towers of ten Unifix blocks, and then halving the number. Ask, 'What do you notice about your answers?' Emphasise that they are the same answer. Repeat with other cards and have students swap roles.

LARGER NUMBERS

- For students who require a challenge, present the equation: 18 × 5. Ask, 'What strategy would you use to solve this problem?' Discuss how the 'multiply by 10 and halve' strategy can be used with larger numbers. Have students solve 18 × 10 to get 180 and then halve their answer to get 90. If necessary, students can model the numbers using NTO 3.7 'Modelling with MAB'. Have students multiply the following numbers by 5: 23, 16, 26, 25, 15, 24, 21 and 17.

REFLECTION

Select from the following to suit your class and their learning outcomes:

- Have students share their work and ask, 'How did multiplying by 10 and halving help you solve the problems quickly and accurately?'
- Present the problem: 7 × 5. Ask, 'What strategy would you use to solve this problem?'

LESSON PLAN 

TUNING IN

BALL PASS

You will need: a ball

Have students stand around the room and tell them that they are going to count by 3s. The student who starts with the ball says '3' and passes the ball to someone else. When the next student catches the ball, they need to say the next number in the pattern. Continue until all students have had a turn.

WHOLE-CLASS INTRODUCTION

NUMBER SEQUENCE

Have students skip count by 3s and record the number sequence on the board. Ask, 'What do you notice?' Discuss patterns, e.g. three numbers with no tens, three numbers with 1 ten and three numbers with 2 tens; or when the digits in numbers are added they have a total of 3, 6 or 9. Ask, 'How would being familiar with this number sequence help us when we multiply numbers by 3?' Present problems such as 5 × 3 and 9 × 3 for students to solve. Ask, 'How did you quickly solve these problems?'

INDEPENDENT TASKS

Note: Choose from Tasks 1, 2 or 3.

You will need: 10-sided dice, counters, NTO 3.11 'Calculator', Student Book p. 77 'Multiply by 3'

TASK 1: NUMBER SEQUENCE COVER-UP

Have students work with a partner and record the counting-by-3s number sequence to 30. Students then take turns to roll a 10-sided dice, multiply that number by 3 and cover their answer on the number sequence with a counter. If the answer is already covered, they miss their turn. The winner is the student who covers the last number in the number sequence. Have students play several times.

TASK 2: INTERACTIVE TASK 

Have students work with a partner. One student selects and records a number from the 3s number sequence, e.g. 24. Their partner then writes the 3 times table fact for that number, e.g. 8 × 3. Have students use NTO 3.11 'Calculator' to check if their problem is correct. If it is, they receive one point. Have students swap roles and continue.

TASK 3: STUDENT BOOK p. 77 *'Multiply by 3'*

TEACHING GROUP

You will need: counters, Unifix blocks

GROUPS OF 3

- For students who require support, have them skip count by 3s and record the number sequence on a large sheet of paper. Students then model each number in the sequence, using counters to make groups of 3. Then have students record the 3 times table fact to match each number. Have students practise memorising the number sequence and the multiplication facts.

DOUBLES AND ADDITION

- For students who require a challenge, explain how we can use doubles facts and addition as a strategy for multiplying numbers by 3. Model the equation 7 × 3 by making three towers of seven Unifix blocks. Separate one tower and emphasise that there are two lots of 7 and one more lot of 7. To find the answer to 7 × 3, we can double the 7 to get 14 and then add one more lot of 7 to get 21. Present other 3 times table facts, e.g. 4 × 3 and 10 × 3, and have students solve them using this strategy.

REFLECTION

Select from the following to suit your class and their learning outcomes:

- Have students share their work and ask, 'How can being familiar with the 3s number sequence help you multiply numbers by 3? Why is it important to have a strategy for multiplying numbers by 3?'
- Have students reflect on Independent Tasks, Task 2. Discuss strategies students used to identify the multiplication fact for the given number. Ask, 'How did you use the 3s number sequence to help you?'

LESSON PLAN 3

TUNING IN

MENTAL STRATEGIES

Have students brainstorm the mental strategies they use when solving multiplication problems. Record them on the board. Ask, 'Why is it important to have mental strategies?'

WHOLE-CLASS INTRODUCTION

MULTIPLICATION 'WHO AM I?'

Explain to students that you are going to 'think aloud' as you solve a multiplication problem and they need to work out what the problem is by thinking of the numbers and the strategy you used. For example: 'I have 6. I will multiply it by 10 to make 60. Now I will halve 60 to make 30 so my answer is 30. What multiplication problem am I?' Discuss strategies students used to determine the problem: 6 × 5. Repeat for other problems, e.g. 'I have 7. I'm going to double it. My answer is 14. What multiplication problem am I?'

INDEPENDENT TASKS

Note: Choose from Tasks 1, 2 or 3.

You will need: BLM 50 '4 in a Row', 10-sided dice, deck of playing cards (only include aces to represent 1s and 2s, 3s, 5s and 10s), counters, BLM 51 'Multiplication Fact Cards', NTO 3.11 'Calculator', Student Book p. 78 'Off to Camp'

TASK 1: 4 IN A ROW

Give pairs of students BLM 50 '4 in a Row' (one copy each). Have them take turns to roll a 10-sided dice and turn over a card from the deck of cards (only include aces to represent 1s and 2s, 3s, 5s and 10s). Have students multiply these numbers together and if possible, cover that number on their game board with a counter. The winner is the first student to have four counters in a row.

TASK 2: INTERACTIVE TASK 

Give pairs of students cards made from BLM 51 'Multiplication Fact Cards'. Place the cards face down and have students take turns to turn over a card. One student solves the problem using a mental strategy; their partner solves it using NTO 3.11 'Calculator'. The student who solves it first wins a point.

TASK 3: STUDENT BOOK p. 78 *'Off to Camp'*

TEACHING GROUP

You will need: BLM 51 'Multiplication Fact Cards'

STRATEGY SORT

- For students who require support, give them cards made from BLM 51 'Multiplication Fact Cards'. Have them solve the problems and then sort the cards into groups based on the mental strategy they used.

READY, SET, MULTIPLY

- For students who require a challenge, have them place cards made from BLM 51 'Multiplication Fact Cards' face down in a pile. Have each student select seven cards, keeping them face down. Say, 'Ready, set, multiply' and have students turn their cards over, solve the problems and record their answers as quickly as possible. The first student to finish and have all answers correct wins that round. Reshuffle the cards and play again.

REFLECTION

Select from the following to suit your class and their learning outcomes:

- Have students reflect on their tasks and ask, 'What mental strategies did you use to quickly and accurately solve the multiplication problems?' Make a list of the strategies and display them in the room for students to refer to during future maths lesson.
- Present the problems: 4 × 2, 8 × 10, 6 × 3, 9 × 1, 3 × 2. Have students explain the mental strategies they would use to solve these problems.

Home Tasks

Select from the possible Home Tasks:

- Have students search through newspapers and magazines and cut out numbers to make the counting-by-3s number sequence.
- Have students practise memorising multiplication facts from the 1, 2, 3, 5 and 10 times tables. Encourage them to use the mental strategies they know.

Assessment

- Have students complete **Student Assessment p. 79**.
- Review with students **Assessment Task Card 3.19**.

During the three lessons:

- Observe which students are able to multiply numbers by 1, 2, 3, 5 and 10 quickly and accurately by using a mental strategy.
- Make note of students who completed the scaffolding tasks or the more challenging activities of the Teaching Groups.
- Review Student Book pages and make notes of areas of difficulty.

Recommendations for Future Learning

Specific to Student Assessment p. 79; if the student is experiencing some difficulty:

Q 1 Review the mental strategy for multiplying numbers by 5. Provide opportunities for the student to multiply numbers by 10 using MAB and then halve their answer. Emphasise that 5 is half of 10 and that is why we can use this strategy.

Q 2 Revise skip counting by 3 and have the student become familiar with the number sequence. Have the student multiply numbers by 3 by modelling groups of 3 and by skip counting by 3s to find the total.

Q 3–4 Revise the mental strategies for multiplying numbers by 1, 2, 3, 5 and 10. Encourage the student to use materials such as counters and MAB to model the strategies. When the student is proficient, have them apply the strategies mentally.

If the student has not achieved the recommended skills for this unit:

1. See **Assessment Task Card 3.19** for specific recommendations.
2. Have the student solve multiplication problems using materials and by becoming familiar with number sequences for counting by 1s, 2s, 3s, 5s and 10s.
3. Review *Nelson Maths: Australian Curriculum NSW Year 2* Unit 23.
4. Have students complete *Building Mental Strategies Skill Book Year 3,* pp. 58–62, to reinforce mental strategies for multiplication.

If the student has achieved the recommended skills and these skills are firmly established, consider:

1. Having the student complete *Building Mental Strategies Skill Book Year 4,* pp. 48–50, to further explore mental strategies for multiplication.
2. Moving forward to *Nelson Maths: Australian Curriculum NSW Year 4* Unit 10.
3. Extending the student by developing mental strategies for multiplying numbers by 6, 8, 7 and 9.

Unit 20 Chance

Statistics and Probability
Chance MA2-19SP describes and compares chance events in social and experimental contexts

certain, even chance, event, fair, impossible, likely, no chance, possible, outcome, unlikely

LESSON PLAN 1

TUNING IN

WHAT DOES IT MEAN?

You will need: book titles or newspaper headlines including the word 'chance'

Present book titles, e.g. *Cloudy with a Chance of Meatballs* (J Barrett, 1978) and *Mostly Sunny with a Chance of Storms* (M Roberts, 2010). Ask, 'What does the word "chance" mean?' Discuss to find out what students know about chance and its reference to the weather. Ask, 'What is the chance that it will be cloudy and rain meatballs?' Discuss the idea that some things have no chance of happening and can be described as impossible. Explain to students that particular terms are used to describe the chance of something happening, and write on the board the words 'impossible', 'certain', 'likely' and 'unlikely'. Have students brainstorm events that they think are impossible, certain, likely and unlikely, and record these.

WHOLE-CLASS INTRODUCTION

HEADS OR TAILS?

You will need: NTO 3.21 'Coin Flip'

Present NTO 3.21 'Coin Flip' to students and ask, 'What can you see? What is on the other side of the coin?' Explain to students that they are going to play a game whereby they need to predict if the coin flip will result in heads or tails. Have students stand and make their prediction by either placing their hands on their head if they think it will be a head or hands on their backside if they think it will be a tail. Flip the coin and those students who predicted correctly make another prediction and the others sit down. Continue until one student remains. Ask, 'When a coin is flipped, how many possible outcomes are there? Is one outcome more likely than the other?' Discuss with students that, when a single coin is flipped, there are two possible outcomes and that the outcome of a head or a tail is 1 out of 2, so both outcomes are equally likely and have an 'even chance' of happening.

INDEPENDENT TASKS

Note: Choose from Tasks 1, 2 or 3.

You will need: LO: *L115 'Slushy sludger: questions'*, Student Book p. 80 'What's the Chance?'

TASK 1: WHAT WILL HAPPEN TODAY?

Working in pairs, students write or draw two things that have an 'impossible' chance of happening, two things that are 'certain' to happen, two things that are 'likely' to happen, two things that are 'unlikely' to happen and two things that have an 'even chance' of happening during a normal school day.

TASK 2: INTERACTIVE TASK

Have students explore LO: *L115 'Slushy sludger: questions'* whereby a vending machine squirts coloured 'slushies' into ice-cream cones. Students work out which 'sludge events' are possible and then choose a matching probability word.

TASK 3: STUDENT BOOK p. 80 *'What's the Chance?'*

TEACHING GROUP

COULD HAPPEN 1

- For students who require support, ask them to think about things that could happen to them today. As a group, make a list, e.g. read a book, draw a picture, eat lunch, play with friends, walk home. Encourage students to think of things that are personally relevant to them. Read through the list and ask, 'Do you think some things are more likely to happen during a day at school than others?' Discuss and have students draw a picture of something they think is likely to happen and something that is possible but unlikely to happen.

COULD HAPPEN 2

- For students who require a challenge, have them work in pairs to write ten things that could happen in the classroom today. Discuss with students how some things are more likely to happen than others. Have students order their list from the least likely to the most likely. Have them share their list and reasoning with the group.

REFLECTION

Select from the following to suit your class and their learning outcomes:

- Have students share their work from the Independent Tasks and ask, 'How did you decide which things were more likely to happen?'
- Ask, 'If I told you it was likely/unlikely that we would do something tomorrow at school, what do you think it could be?' Have students give their suggestions and discuss their reasoning.

LESSON PLAN 2

TUNING IN

WHEN PIGS HAVE WINGS

Explain that there are many sayings that have something to do with chance. Write 'It will happen when pigs have wings' on the board, and ask students if they know what it means. Discuss that it means there is no chance or it is impossible. Ask, 'If I roll a dice with the numbers 1 to 6, what is my chance of rolling a 7?' Discuss possible outcomes and how there is no chance at all of rolling 7. Have students think of other things that there is no chance of happening. Record their ideas.

WHOLE-CLASS INTRODUCTION

LUCKY THREE

You will need: NTO 3.3 'Six-Sided Dice'

Present NTO 3.3 'Six-Sided Dice' and ask, 'If I roll the dice, what number could I get?' Have students name the numbers that could be rolled. Tell students that you would like to roll a 3 because that is your favourite number. Ask, 'Could I roll 3 next? Do you think it is likely that I will roll 3?' Discuss with students that, while it is possible to roll 3 as it is one of the numbers on the dice, the chance of rolling 3 is 1 out of 6. Have students predict how many times they think the dice will need to be rolled before 3 comes up. Use the NTO and roll until 3 comes up. Ask, 'If I want to roll 3 again, do you think it will take the same number of rolls? Why?'

INDEPENDENT TASKS

Note: Choose from Tasks 1, 2 or 3.

You will need: BLM 52 'Car Race', card, yellow and red pencils, BLM 53 'Spinners', paperclips, sharp pencils, LO: *L2378 'Spinners: predict and test'*, Student Book p. 81 'Spinners'

TASK 1: RACING CARS

Using BLM 52 'Car Race', students work in pairs to cut out and paste the first race track onto card. They then cut out and colour the two cars, one yellow and one red. Using the first spinner on BLM 53 'Spinners', students colour one section red and two sections yellow. Each student chooses a car and places it on 'Start' under the first of the column of the race track. They take turns to spin the spinner. The yellow car needs to land on yellow to move one square; the red car needs to land on red to move one square. Have students predict which car they think will win the race. Then conduct the race. When all races are complete, have students share their results. Have students look at the final positions of each car and discuss variations in results. Have students design and colour a spinner so that both cars have an even chance of winning. Students can use the other track to conduct a race to test their spinner.

TASK 2: INTERACTIVE TASK 

Using LO: *L2378 'Spinners: predict and test'*, students race two cars along a track determined by the results of a spinner. Students compare actual results with predicted results.

TASK 3: STUDENT BOOK p. 81 *'Spinners'*

TEACHING GROUP

You will need: NTO 3.22 'Spinner', BLM 53 'Spinners', paperclips, sharp pencils, small whiteboards or paper, dice

SPIN THE SPINNER

- For students who require support, have them explore spinners so that they can make the connection between the area on the spinner and the likelihood of occurrence. Present students with NTO 3.22 'Spinner' and have all segments the same colour. Ask, 'When I spin the spinner, what colour do you think it will land on? Why?' Discuss how, as the whole area of the spinner is one colour, it can only land on that colour. Invite all students to spin the spinner. Next, click on one segment and ask, 'When I spin the spinner, what colour could it land on? Which colour do you think it will land on? Why?' Invite students to spin the spinner. Continue to click on segments, discuss possible outcomes and determine most likely outcomes and test.

SPINNING CHANCE

- For students who require a challenge, give pairs of students a copy of BLM 53 'Spinners' and have them make the fourth spinner. Students roll a dice and record the numbers on the segments of the spinner. Have students write a statement about the likelihood of landing on each number on their spinner. Pairs of students then swap statements with another pair and use the information to make an identical spinner. Have students compare spinners and ask, 'Are they the same? Why?'

REFLECTION

Select from the following to suit your class and their learning outcomes:

- Have students share their work from the Independent Tasks and the Teaching Groups, and discuss the findings from their chance experiments in the Independent Tasks. Ask, 'Did all outcomes have the same chance of happening? Why?'
- Present NTO 3.22 'Spinner' and click on five segments to change the colour. Ask, 'Which colour are we more likely to land on? How do you know?' Discuss with students how they can calculate the chance by counting the segments of same colour and comparing to the total number of segments in the circle. Change the colours on the spinner and ask, 'What are the chances of landing on each of the colours now?'

LESSON PLAN 3

TUNING IN

WHICH LETTER?

You will need: a 10-sided dice

Take students to an area where they can line up side by side. Give each student a letter of the alphabet. Tell them that you are going to roll a 10-sided dice, and if the number rolled has their letter in it, then they can take a step forward. The first student to take five steps is the winner. After a student has won the game, ask, 'Did everyone have the same chance of winning? Why?' Discuss that students with the letters a, b, c, d, j, k, l, m, p, q, y and z have no chance of moving forward as no number uses those letters, and that as the letter e is used in seven numbers, the student with the letter 'e' would be the most likely to win. Ask, 'If we played the game again, do you think we would get the same result? Why?'

WHOLE-CLASS INTRODUCTION

IN THE BAG

You will need: a paper bag, counters

Place a red, blue, green and yellow counter into the paper bag and ask, 'If I take a counter from the bag, what colour might it be?' Tell students that your favourite colour is red and ask, 'What is my chance of taking a red counter from the bag?' Discuss that red is one of four counters in the bag, so the chance is one in four. Shake the bag, select a counter from the bag and place it back in the bag until a red counter has been collected. Repeat. Ask, 'How can I increase my chance of selecting a red counter from the bag?' As students make suggestions (e.g. add more red counters, place only two counters in), have them work out the chance. Trial students' suggestions.

INDEPENDENT TASKS

Note: Choose from Tasks 1, 2 or 3.

You will need: BLM 4 'Blank Cards' with a W, I or N written on the cards; paper bags; LO: *L116 'The slushy sludger: best guess'*; Student Book p. 82 'Chance Machines'

TASK 1: WIN

Students can play with a partner or in a small group. Each group needs a paper bag and three cards with the letters W, I and N, which are placed in the bag. Each student takes it in turns to shake the bag and select a card. They record the letter and place the card back in the bag. When a student has recorded at least one of each letter, they tally how many times they had to draw a card from the bag to make the word 'WIN'. Have students share how many times it took. Ask, 'What was the least and the most number of draws?' Discuss the variation in results.

TASK 2: INTERACTIVE TASK 

Students can explore LO: *L116 'The slushy sludger: best guess'* whereby a vending machine squirts 'slushies' into ice-cream cones. The machine serves coloured slush randomly from four slots, and students work out which colour is the most common (most likely to be served). However, least common colours are sometimes served. After several guesses, students can check their results from the random sample.

TASK 3: STUDENT BOOK p. 82 *'Chance Machines'*

TEACHING GROUP

You will need: paper bags, coloured counters, several paper bags containing different colour combinations of eight counters

HALF AND HALF

- For students who require support, have them work with a partner to put an equal number of two different coloured counters into a paper bag. Ask, 'If you were to draw a counter 20 times from the bag, how many of each colour do you think you would draw?' Discuss students' reasoning for their predictions. Have them shake the bag, draw a counter, record its colour and place it back into the bag. Repeat the process 20 times. Have students share their results and discuss in relation to their predictions.

WHAT'S IN THE BAG?

- For students who require a challenge, have them work with a partner to draw a counter from a paper bag containing a different combination of eight counters. They then place the counter back into the bag and repeat this process 20 times. Have students record their results, and then based on their results, make a prediction about what they think is in the bag. Have students share predictions and discuss their reasoning.

REFLECTION

Select from the following to suit your class and their learning outcomes:

- Have students share and disuss the results of their chance activities. Ask, 'Were your results always the same? Why or why not?'
- In front of students, place a blue, green and yellow counter and four red counters into a paper bag. Ask, 'What colour counter am I most likely to draw out of the bag? Does that mean I **will** draw out a red counter?' Have students explain their reasoning.

Home Tasks

Select from the possible Home Tasks:

- Have students think about events that could happen at home in the next week: one event that is certain to happen, one that is unlikely to happen, one that has an even chance of happening and one that is unlikely to happen.
- Have students teach someone at home how to play WIN, from Lesson Plan 3 , Independent Tasks, Task 1.

Assessment

- Have students complete **Student Assessment p. 83**.
- Review with students **Assessment Task Card 3.20**.

During the three lessons:

- Collect created items from Lesson Plan 1, Independent Tasks, Task 1, as work samples for students' portfolios.
- Make note of students completing the scaffolding tasks or the more challenging activities of the Teaching Groups.
- Review Student Book pages and make notes of areas of difficulty.

Recommendations for Future Learning

Specific to Student Assessment p. 83; if the student is experiencing difficulty:

Q 1 Encourage the student to use the language of chance by having them predict the likelihood of a particular event happening each day, e.g. the principal coming into the room, the school being flooded, a student from another class entering the classroom. Check the student's predictions before they leave at the end of the day.

Q 2 Have the student use spinners to play games and use NTO 3.22 'Spinner' to explore the results of shading the spinner in different ways.

Q 3 Look for opportunities to conduct chance experiments, e.g. drawing counters from a bag to group students randomly or holding reward raffles where students can earn tickets.

If the student has not achieved the recommended skills for this unit:

1. See **Assessment Task Card 3.20** for specific recommendations.
2. Review *Nelson Maths: Australian Curriculum NSW Year 2* Unit 29.

If the student has achieved the recommended skills and these skills are firmly established, consider:

1. Having the student make their own chance boardgame using a coloured spinner to move players around the board. Have them explore the concept of fairness in relation to how the game has been designed.
2. Moving forward to *Nelson Maths: Australian Curriculum NSW Year 4* Unit 19.

Unit 21 Patterns

Number and Algebra

Patterns and algebra MA2-8NA generalises properties of odd and even numbers, generates number patterns, and completes simple number sentences by calculating missing values

100 chart, 2-step number pattern, add on, continue, number line, number pattern, rule, subtract

LESSON PLAN 1

TUNING IN

ACTION PATTERNS

Have students sit in a circle. Start an action pattern by clapping hands twice then patting legs twice (clap, clap, pat, pat, clap, clap etc.). Have students follow the pattern. Stop, then ask, 'What were we doing?' Discuss making a pattern with actions. Select one student to devise and start a new action pattern. Once other students have worked out the pattern, they join in. Limit each pattern to three types of movements (e.g. clap, click and stomp) – the number of each is up to the student. Once all students have joined in, stop and discuss. Repeat with two or three other students selecting the action pattern.

WHOLE-CLASS INTRODUCTION

EXPLORING PATTERNS IN A 100 CHART

You will need: NTO 3.9 '100 Chart'

Display NTO 3.9 '100 Chart' with 'show numbers' selected. Explain that the chart includes all numbers from 1 to 100, adding '1' as you move across the chart, and adding '10' as you move down. Reset the screen and then highlight every third square to about halfway down the chart (using a single click on each box). Ask, 'Can you see a pattern? How would you describe it?' Show the numbers on the chart. Ask, 'Can you see the number pattern? What number would come next?' Have students make predictions, and select students to highlight squares. Ask, 'How did you know which number would be next in the pattern? What is the rule for this pattern?' Discuss the pattern of adding 3. Select students to finish the pattern by highlighting the numbers. Repeat for other counting patterns.

INDEPENDENT TASKS

Note: Choose from Tasks 1, 2 or 3.

You will need: BLM 14 '100 Chart', dice, LO: *L589 'Musical number patterns: music maker'*, Student Book p. 84 'Jigsaw 100 Chart'

TASK 1: PATTERNS ON A 100 CHART

Give each student BLM 14 '100 Chart' and a dice. Students roll the dice and count by this number on the 100 chart, shading each square. Have them describe the pattern they have made on the chart. Roll again and repeat for another number, using a different colour to shade numbers. Describe the pattern and any similarities or differences to the previous one. Repeat for one or two other numbers.

TASK 2: INTERACTIVE TASK 

Have students work independently on computers to play LO: *L589 'Musical number patterns: music maker'*. Students create a rule by selecting a starting number and a 'count by' number. They then model the pattern on the number line and listen to the musical notes.

TASK 3: STUDENT BOOK p. 84 *'Jigsaw 100 Chart'*

TEACHING GROUP

You will need: BLM 3 'Blank Chart', different coloured highlighters, transparent copies of BLM 14 '100 Chart'

100-CHART PATTERNS

- For students who require support, give them BLM 3 'Blank Chart' and a highlighter. Have students count every fifth square and draw a cross in them. After a few lines, ask them to look for the pattern, then use the pattern to fill remaining squares. Have them use a different coloured highlighter and count by 10s, drawing a circle in each square. Look at the pattern and discuss. Then, overlay the transparent copy of BLM 14 '100 Chart'. Discuss the numbers that are highlighted and look for patterns.

JIGSAW MAKER

- For students who require a challenge, give them two copies of BLM 3 'Blank Chart' and have them make a number chart jigsaw. Students use one copy of BLM 3 to devise their number chart – it may start and finish at any number and may count by any number between zero and 5. The second copy of BLM 3 will be their jigsaw board. Have them first shape the jigsaw pieces, then fill in some of the numbers on each piece. Cut the pieces out and swap with a partner to solve.

REFLECTION

Select from the following to suit your class and their learning outcomes:

- As a class, play LO: *L589 'Musical number patterns: music maker'*.
- Have students share their work from the Independent Tasks, and ask 'What did you discover about patterns on the 100 chart?'

LESSON PLAN 2

TUNING IN

DOT PATTERNS

Draw the following sequence on the whiteboard:

Ask, 'What comes next?' Have them copy the sequence into their workbooks and draw four more possible steps. Compare responses. Ask, 'Is it possible to have more than one correct response?' Draw this further along the board:

Ask, 'What if this is step 7 – how many correct responses are there now?'

WHOLE-CLASS INTRODUCTION

PATTERNS ON A NUMBER LINE

You will need: NTO 3.6 'Number Line'

Display NTO 3.6 'Number Line'. Highlight every fourth mark on the number line. Ask, 'Can you explain the pattern that has been made? If the first mark on the number line is 1, what would the second mark be?' Select a student to write the numbers using the IWB marker and identify the pattern. Erase the numbers and ask, 'If the first number on the line is 3, what would the pattern be?' Select a student to renumber the number line. Discuss the new number pattern. Repeat with different starting numbers and different number patterns.

INDEPENDENT TASKS

Note: Choose from Tasks 1, 2 or 3.

You will need: BLM 55 'Start at... Cards', dice, LO: *L1084 'Hopper: whole numbers'*, Student Book p. 85 'Bounce, Bounce, Stop!'

TASK 1: NUMBER LINE CHALLENGES

Have students work in pairs with BLM 55 'Start at... Cards' and a dice. Students are to create number line challenges for each other by randomly selecting a card, then rolling the dice and deciding whether their partner is to add or subtract this number (if the starting number is small, add; if large, subtract). Each completed number line should contain ten numbers. Have students discuss patterns with their partner and continue to take turns in creating number line challenges.

TASK 2: INTERACTIVE TASK 

Have students work independently on computers, using LO: *L1084 'Hopper: whole numbers'* to skip count and predict an end point on a number line. They also examine patterns formed on the number chart.

TASK 3: STUDENT BOOK p. 85 *'Bounce, Bounce, Stop!'*

TEACHING GROUP

You will need: dice, counters, BLM 14 '100 Chart', A3 sheet of paper, BLM 57 'Chinese Zodiac'

USING THE 100 CHART TO UNDERSTAND NUMBER LINES

- For students who require support, have them roll a dice and use the number rolled and counters to map out the number pattern on BLM 14 '100 Chart'. Then, draw a number line on a landscape sheet of A3 paper and transfer the numbers from the 100 chart to the number line. Ask, 'We started at zero – where would I write this? Next was ... – where would this go?' and so on. Draw out links between the 100 chart and the number line. Discuss any patterns formed.

CHINESE ZODIAC

- For students who require a challenge, give them BLM 57 'Chinese Zodiac'. Explain that the Chinese zodiac follows a 12-year cycle, and have them create number patterns by adding 12 to each year shown until they find the year they were born – their Chinese zodiac sign! They may then look for the Chinese zodiac signs of their parents, siblings or friends.

REFLECTION

Select from the following to suit your class and their learning outcomes:

- Invite students to share some of the patterns they found in their Independent Tasks or Teaching Groups. Ask, 'When were the patterns challenging?'
- Have students stand in a circle, give them a rule for a number pattern, e.g. 'Starting at 6, add 4', and have students continue the pattern around the circle.

LESSON PLAN 3

TUNING IN

CREATE A PIN

Have students create a 4-digit PIN that is easy to remember. Students then explain why they chose it and how they will remember it. Then have students devise a 6-digit PIN. Ask, 'Did you use the same or a different idea as for your 4-digit PIN? Why?'

WHOLE-CLASS INTRODUCTION

2-STEP RULES

You will need: BLM 55 'Start at... Cards', BLM 56 'Number Pattern Rules'

Write on the board: 3, 7. Have students predict the next number. Most will predict that the rule is 'add 4', and answer 11. Ask, 'Could there be another answer?' Discuss ideas. Write the next two numbers as 8, 12. Ask, 'What could the rule be?' If needed, write the next two numbers: 13, 17. The rule is '+ 1, + 4'. This is a 2-step rule. Have one student select a card from BLM 55 'Start at... Cards' and two cards from BLM 56 'Number Pattern Rules'. Select the order to perform the rules and work with students to create a number pattern, e.g. start at 10, add 5 then subtract 2 = 10, 15, 13, 18, etc.

INDEPENDENT TASKS

Note: Choose from Tasks 1, 2 or 3.

You will need: BLM 55 'Start at... Cards', BLM 56 'Number Pattern Rules', *PowerPoint,* Student Book p. 86 '2-Step Number Patterns'

TASK 1: RULE SOLVERS

Give pairs of students cards made from BLM 55 'Start at... Cards' and BLM 56 'Number Pattern Rules'. Have Player A randomly select one card from BLM 55 and two from BLM 56 to create a rule. Keep them hidden from Player B. Player A begins writing numbers following the rule, giving time for Player B to guess the rule. Once solved, count the numbers written by Player A – this is Player B's score. Then Player B writes a sequence for Player A to guess. Continue, taking turns to play. The winner is the player with the lowest score (who took fewer guesses to find the rules).

TASK 2: INTERACTIVE TASK 

Give students a card from BLM 55 'Start at... Cards' and two from BLM 56 'Number Pattern Rules' and use them to create a rule. Have them create a *PowerPoint* presentation of their number sequence. The sequence should include 20 numbers, with the final slide containing the rule. If time, students can devise their own rule to create another *PowerPoint* presentation.

TASK 3: STUDENT BOOK p. 86 *'2-Step Number Patterns'*

TEACHING GROUP

You will need: BLM 55 'Start at... Cards', BLM 56 'Number Pattern Rules', BLM 14 '100 Chart'

ONE STEP AT A TIME

- For students who require support, use one card from BLM 55 'Start at... Cards' and an addition card from BLM 56 'Number Pattern Rules' to create a 1-step rule. Give each student BLM 14 '100 Chart' to mark their starting point, then mark squares according to the rule. Once they have about ten numbers, students write the sequence in their workbooks. Once consolidated, try a 2-step rule using two addition cards.

TWO STEPS IN ONE

- For students who require a challenge, write 2, 5, 11, 23 on the board. Ask, 'What could the rule be?' Discuss. This pattern shows two steps as a single move – double, add 1. Have students devise a series of two-steps-in-one patterns to challenge a partner. They may use multiplication or division.

REFLECTION

Select from the following to suit your class and their learning outcomes:

- Write the sequence 2, 4, 2, 4, 2, 4 on the board. Ask, 'What 2-step rule could this pattern be following?' (start at 2, double then subtract 2)
- Invite students who completed Independent Tasks, Task 2, to show their *PowerPoint* presentations.

Home Tasks

Select from the possible Home Tasks:

- Have students use one month on a grid-style calendar to create three different pattern rules. The pattern should be shaded on the calendar and the rules written below.
- Give students a counting rule to work on at home, creating a pattern of 20 numbers, and presenting it in an interesting way, e.g. threaded on a piece of string.

Assessment

- Have students complete **Student Assessment p. 87**.
- Review with students **Assessment Task Card 3.21**.

During the three lessons:

- Collect examples of students' 100 charts and number lines from Lesson Plans 1 and 2 to add to their portfolios as evidence of their ability to understand patterns.
- Make note of how students go about solving the 2-step number patterns in Lesson Plan 3, Independent Tasks, as evidence of their ability to solve more complex patterns.
- Review Student Book pages and make notes of areas of difficulty.

Recommendations for Future Learning

Specific to Student Assessment p. 87; if the student is experiencing some difficulty:

Q 1 Revise the 100 chart, modelling how to count by 2s, 3s and 4s and shading each square.

Q 2 Revisit the 'Patterns on a Number Line' activity from the Whole-Class Introduction in Lesson Plan 2.

Q 3 Revise the 2-step number pattern activities from Lesson Plan 3. Simplify the process by providing a 100 chart or number line, and shading/marking a 2-step addition pattern.

If the student has not achieved the recommended skills for this unit:

1. See **Assessment Task Card 3.21** for specific recommendations.
2. Have the student work with basic 1-step number patterns before moving on to more complex ones.
3. Work with number patterns using addition. Once consolidated, move on to subtraction.
4. Use calculators or model with concrete materials to aid student understanding.
5. Review *Nelson Maths: Australian Curriculum NSW Year 2* Units 18 and 19.

If the student has achieved the recommended skills and these skills are firmly established, consider:

1. Having the student complete *Building Mental Strategies Skill Book Year 3*, pp. 4–5.
2. Moving forward to *Nelson Maths: Australian Curriculum NSW Year 4* Unit 21.
3. Extending the student to make number patterns using fractions and decimals.

Unit 22 Multiplication

Number and Algebra
Multiplication and division MA2-6NA uses mental and informal written strategies for multiplication and division

array, groups of, multiplication, multiply, rows of, worded problem

LESSON PLAN 1

TUNING IN

CHOCOLATE BAR ARRAY

Say, 'I had a chocolate bar with three rows and five squares of chocolate in each row.' Have students discuss what the chocolate bar would look like. Have a student draw the chocolate bar on the board. Establish that it is an array with three rows of 5. Ask, 'How many pieces of chocolate were there altogether? What multiplication problem does this represent?'

WHOLE-CLASS INTRODUCTION

ARRAY GAME

Record the problem 4 × 2 on the board. Explain to students that when they hear the word 'array', they need to form a group and sit as an array to match the problem on the board. Have them walk around the room. Say 'array' and ensure students are sitting in four rows of 2. If a student is not part of an array, they are out of the game. Record the problem 2 × 4 and have students reorganise themselves to match the new equation. Ask, 'What do you notice about the two arrays?' Repeat by recording a different multiplication problem on the board. The winners are the final students left in the game.

INDEPENDENT TASKS

Note: Choose from Tasks 1, 2 or 3.

You will need: dice, BLM 10 'Grid Paper', LO: *L106 'The array'*, Student Book p. 88 'Sports Room Arrays'

TASK 1: ARRAY COVER-UP

Have students play with a partner, taking turns to roll two dice and using the numbers rolled to form a multiplication problem. Students then colour an array to match on BLM 10 'Grid Paper'. If they are unable to fit their array on the grid paper, they miss their turn. The winner is the student who colours the final square on the grid paper.

TASK 2: INTERACTIVE TASK 

Have students use LO: *L106 'The array'* to explore the connection between arrays and multiplication problems.

TASK 3: STUDENT BOOK p. 88 ***'Sports Room Arrays'***

TEACHING GROUP

You will need: dice, BLM 4 'Blank Cards' (multiple copies), BLM 5 'Digit Cards', butcher's paper

TWO ARRAYS

- For students who require support, roll two dice and form a multiplication problem, e.g. 6 × 4. Have students model the equation using multiple cards from BLM 4 'Blank Cards' to form an array. Rearrange the dice to form another problem, e.g. 4 × 6, and have students model this equation. Discuss how the array has been rotated and both have the same product. Have students roll two dice and then arrange the blank cards to identify the two possible arrays.

GUESS THE ARRAY

- For students who require a challenge, have them work with a partner. Have one student cover their eyes while their partner turns over two cards from BLM 5 'Digit Cards', draws a large array on a sheet of butchers' paper and places their hand over the array to cover parts of it. The first student then looks at the array and tries to work out the multiplication problem and the product. Students then swap roles.

REFLECTION

Select from the following to suit your class and their learning outcomes:

- Have students explain their understandings of multiplication and arrays. Encourage them to use examples from their work to aid their explanation.
- Have students reflect on the 'Guess the Array' task in the Teaching Group. Ask, 'What strategies did you use to work out the multiplication problem and the product?'

LESSON PLAN 2

TUNING IN

MULTIPLICATION PAIRS

You will need: BLM 5 'Digit Cards'

Give each student a card made from BLM 5 'Digit Cards'. Say a number, e.g. 20, and have students walk around the room to find someone holding a card that makes 20 when multiplied with their card, e.g. a student holding 2 could pair with a student holding 10 because 10 × 2 = 20. When they make a pair, the students sit down together. The first pair to sit down receives a point. Repeat with other numbers, e.g. 18, 9, 12.

WHOLE-CLASS INTRODUCTION

MULTPLYING LARGER NUMBERS

You will need: NTO 3.7 'Modelling with MAB'

Present the problem: 13 × 3. Ask, 'How could we solve this problem?' Discuss how we need to work out what three lots of 13 equals. Establish that 13 is made of 1 ten and 3 ones. Use NTO 3.7 'Modelling with MAB' to model 13 × 3 by setting out 1 ten and 3 ones three times, to form an array. Have students count up the ones and then the tens to determine the answer of 39. Then present the problem: 16 × 4. Model solving the problem by setting out 1 ten and 6 ones four times, to form an array. Have students count up the ones. Ask, 'What should we do now that there are 24 ones?' Discuss students' ideas. Model regrouping the 24 ones as 2 tens and 4 ones. Then have students count up the tens to determine the answer of 64. Repeat with other problems, e.g. 12 × 4, 24 × 3 and 18 × 6.

INDEPENDENT TASKS

Note: Choose from Tasks 1, 2 or 3.

You will need: BLM 12 'Sports Equipment Prices', MAB, BLM 1 '2-Digit Number Cards', dice, NTO 3.7 'Modelling with MAB', Student Book p. 89 'Multiplication Puzzle'

TASK 1: AT THE SPORTS SHOP

Give each student BLM 12 'Sports Equipment Prices'. Explain that they need to buy the following items: skipping rope, cricket ball, goggles, netball, football, tennis racquet, basketball, soccer ball and hockey stick. Have them select items and calculate how much it would cost to buy two, three, four or five of them. Have students use MAB to model and solve the problems.

TASK 2: INTERACTIVE TASK 

Give students cards made from BLM 1 '2-Digit Number Cards' and a dice. Place the cards face down. Have students turn over a card and roll the dice to form a multiplication problem, e.g. 48 × 3. Have students model and solve the problem using NTO 3.7 'Modelling with MAB'.

TASK 3: STUDENT BOOK p. 89 *'Multiplication Puzzle'*

TEACHING GROUP

You will need: BLM 58 'Multiplication Cards', MAB, BLM 5 'Digit Cards'

SORT THEM OUT

- For students who require support, give them cards from BLM 58 'Multiplication Cards'. Have them solve the problems using MAB to model the lots of tens and ones to form an array. Support students with regrouping where necessary. Have students then sort the problems into two categories: those that required regrouping and those that did not.

IN BETWEEN

- For students who require a challenge, explain that they need to make multiplication problems that have an answer between 200 and 400. Give them cards made from BLM 5 'Digit Cards'. Have them select three cards to make a 2-digit number and a 1-digit number, e.g. with cards 5, 6 and 2, the numbers 56 and 2 could be made. Have students multiply their numbers together by modelling them with MAB. If the answer is between 200 and 400, have them record the problem.

REFLECTION

Select from the following to suit your class and their learning outcomes:

- Have students share their work on multiplying larger numbers and explain the strategies they used. Ask, 'How do you know when a multiplication problem requires regrouping?'
- Have students present the problems they recorded in the 'In Between' Teaching Group activity. Ask, 'What strategies did you use to find problems that had an answer between 200 and 400?'

LESSON PLAN

TUNING IN

PENCILS AND CUPS

You will need: pencils and cups

Show students four cups with five pencils in each cup. Ask, 'What multiplication problem can you see?' Discuss how there are four lots of five pencils and record the equation: 4 × 5 = 20. Have students discuss the worded problem that would match this equation. Record a problem, e.g. 'There are 4 cups with 5 pencils in each cup. Altogether there are 20 pencils.' Then present the problem: 'There are 5 cups with 12 pencils in each cup. How many pencils are there?' Ask, 'How would we solve this problem?'

WHOLE-CLASS INTRODUCTION

VERTICAL RECORDING

You will need: NTO 3.7 'Modelling with MAB'

Present the equation: 23 × 3 (horizontally). Explain that this equation can be recorded in another way. Record the equation vertically and discuss how it is asking us to work out three lots of 23. Discuss how we first multiply the 3 ones by 3 and then the 2 tens by 3. If necessary, use NTO 3.7 'Modelling with MAB' to model. Model how the answer is recorded. Emphasise how the ones and tens in the problem are lined up vertically. Then present the equation: 27 × 2 (horizontally). Have a student demonstrate how the problem is recorded vertically. Multiply the 7 ones by 2 and ask, 'What should we do with the 14 ones?' Discuss how the 14 ones are regrouped as 1 ten and 4 ones. Model how this is recorded on the algorithm. Multiply the 2 tens by 2 and add the extra ten that was regrouped and record the answer. Repeat with other equations, e.g. 35 × 2 and 14 × 5.

INDEPENDENT TASKS

Note: Choose from Tasks 1, 2 or 3.

You will need: BLM 59 'Multiplication Pictures', deck of playing cards (picture, joker and number 10 cards removed), NTO 3.11 'Calculator', Student Book p. 90 'Multiplication Problems'

TASK 1: WRITE AND SOLVE A PROBLEM

Give each student BLM 59 'Multiplication Pictures'. Have them write a worded problem to match each picture, then solve the problem by vertically recording the number problem.

TASK 2: INTERACTIVE TASK 

Have students turn over three cards from a deck (picture, joker and number 10 cards removed), form a 2-digit number and a 1-digit number, e.g. 34 and 2, and form a multiplication problem, e.g. 34 × 2. Students record the problem vertically, then solve it. Have students check their answer using NTO 3.11 'Calculator'. They can then write worded problems to match.

TASK 3: STUDENT BOOK p. 90 *'Multiplication Problems'*

TEACHING GROUP

You will need: BLM 60 'Think Board', calculators, LO: *L61 'The multiplier: make your own easy multiplications'*

THINK BOARDS

- For students who require support, have them represent multiplication problems in different ways using BLM 60 'Think Board'. For the equation 12 × 3, have them record the number sentence, a worded problem, a picture and a representation of the materials. Encourage students to record the equation vertically next to the number sentence. Have students use a calculator to check their answers. If their answer is correct, have them draw a star next to the number sentence. Repeat with other equations, e.g. 9 × 4, 11 × 3, 15 × 3.

THE MULTIPLIER

- For students who require a challenge, have them solve multiplication problems using LO: *L61 'The multiplier: make your own easy multiplications'*. Discuss the strategies they used to solve the problems.

REFLECTION

Select from the following to suit your class and their learning outcomes:

- Present students with the equation: 9×6. Have them use the cups and pencils from the Tuning In activity to model this equation.
- Have students explain how to solve and record vertical multiplication problems. Ask, 'How do you know if you need to regroup ones as a ten when solving a multiplication problem?'

Home Tasks

Select from the possible Home Tasks:

- Have students write worded multiplication problems to match situations around the home, e.g. 'There are 4 bowls with 2 scoops of ice-cream in each bowl.'
- Have students search for arrays around the home, e.g. 2 rows of 6 eggs in an egg carton.
- Have students identify 2-digit numbers in magazines and multiply these numbers by 2, 3 or 5.

Assessment

- Have students complete **Student Assessment p. 91**.
- Review with students **Assessment Task Card 3.22**.

During the three lessons:

- Observe which students are able to solve multiplication problems using arrays. Identify students who are able to multiply 2-digit numbers by 1-digit numbers.
- Make note of students who completed the scaffolding tasks or the more challenging activities of the Teaching Groups.
- Review Student Book pages and make notes of areas of difficulty.

Recommendations for Future Learning

Specific to Student Assessment p. 91; if the student is experiencing some difficulty:

Q 1 Revise arrays and how they represent multiplication problems. Present the student with arrays and have them write the matching multiplication problems. Present them with multiplication problems and have them use counters to model the arrays.

Q 2–3 Review how to multiply 2-digit numbers by 1-digit numbers. Have the student use MAB to represent multiplication problems by modelling the tens and ones and arranging them to form an array. Revise the process of regrouping ones as tens when necessary.

Q 4–5 Revise how to solve worded multiplication problems. Discuss strategies for identifying the multiplication problem and solving it.

If the student has not achieved the recommended skills for this unit:

1. See **Assessment Task Card 3.22** for specific recommendations.
2. Have the student use materials such as MAB or counters to model arrays and multiplication problems. Review *Nelson Maths: Australian Curriculum NSW Year 2* Unit 23.
3. Have the student complete *Building Mental Strategies Skill Book Year 3*, pp. 58–62, to reinforce mental strategies for multiplication.

If the student has achieved the recommended skills and these skills are firmly established, consider:

1. Having the student complete *Building Mental Strategies Skill Book Year 4*, pp. 48–50, to further explore mental strategies for multiplication.
2. Moving forward to *Nelson Maths: Australian Curriculum NSW Year 4* Unit 16.
3. Extending the student by introducing multiplication involving larger numbers.

Unit 23 Area

Measurement and Geometry

Area MA2-10MG measures, records, compares and estimates areas using square centimetres and square metres

area, length, measurement units, space covered, square centimetre (cm^2), square metre (m^2), width

LESSON PLAN 1

TUNING IN

WHAT IS AREA?

Write the word 'Area' on the board. Have students copy this in the middle of a blank page in their workbooks and brainstorm everything they know about area. Share ideas to create a class brainstorm on the board. Promote student understanding that area is the size of a surface.

WHOLE-CLASS INTRODUCTION

USING INFORMAL MEASURES

You will need: NTO 3.23 'Area of Shapes'

Display NTO 3.23 'Area of Shapes' and select a rectangle from the shape options. Discuss what area means in relation to this shape, and look at the 'informal measures' options. Move one leaf onto the shape and say, 'I'd like to measure the area of my shape in leaves. How many leaves do you think I would need to cover this shape?' Discuss. Move leaves and attempt to cover the shape, making sure they do not overlap. Discuss the effectiveness of leaves to measure area and ask, 'Do they cover the whole shape?' Highlight the need for good measures to be accurate, consistent and reproducible. Turn on the grid. Ask, 'How can the squares on the grid help us to estimate area?' Select another option from the list and test it. Discuss.

INDEPENDENT TASKS

Note: Choose from Tasks 1, 2 or 3.

You will need: classroom items (e.g. books, calculators, playing cards, counters), *Word,* NTO 3.23 'Area of Shapes', BLM 10 'Grid Paper', Student Book p. 92 'Area'

TASK 1: AREA OF A CALCULATOR

Have students work in pairs to find the area of a calculator using informal classroom items, such as counters. Have students estimate the area first, then measure the area of the calculator with their selected items. Next remind students of the usefulness of the grid squares on NTO 3.23 'Grid Paper' when estimating area. Identify the squares on the grid paper as square centimetres and introduce the square centimetre as a formal unit for measuring area. Have the students place the calculator on the centimetre square grid paper to find its area in square centimetres. Introduce cm^2 as the abbreviation for square centimetres.

TASK 2: INTERACTIVE TASK 

Have students work independently on computers, using NTO 3.23 'Area of Shapes' to experiment with using informal measures to find the area of different shapes. On completion, have them write five 'helpful hints' for measuring the area of a shape. These could be typed using Word.

TASK 3: STUDENT BOOK p. 92 *'Area'*

TEACHING GROUP

You will need: classroom items (e.g. workbooks, sticky notes, CD), newspapers, chalk

AREA OF A BOOK

- For students who require support, select a smaller item to measure, e.g. their workbook. Hold up a sticky note and ask, 'How many sticky notes would it take to cover your book, without overlapping?' Have students make predictions then cover the area of the book with sticky notes. Have them record the area of their book in sticky notes. Try other objects in the room and record results. Show them a CD or circular object and discuss its effectiveness as a measure.

AREA OF THE CLASSROOM

- For students who require a challenge, have them work in groups and attempt to find the area of the classroom in newspaper sheets. Students draw a diagram of the classroom and add details of their measuring to assist them. They may use chalk to create small markers on the floor to track their progress. Make sure their newspaper sheets are facing the same way. Once some progress has been made, have them brainstorm how they could complete the task mathematically.

REFLECTION

Select from the following to suit your class and their learning outcomes:

- Have students share any interesting findings from the Independent Tasks, Tasks 1 and 2.
- Ask, 'If I was going to get a new cover for my table that needed to fit the area exactly, what problems could I encounter if I gave the measure in calculators, dictionaries or hands?'

LESSON PLAN 

TUNING IN

PADDOCKS

You will need: square dot paper on the IWB (this is readily available on IWB software), BLM 62 'Square Dot Paper' cut into quarters

Using square dot paper on the IWB, model how to play 'Paddocks'. Players take turns to put a line either vertically or horizontally between two lines. Once a square has three sides, the next player may put the fourth line in and claim that square by writing their initial in it. After claiming a square, they get another turn. The winner is the player to have the most claimed squares. Give pairs of students a quarter of BLM 62 'Square Dot Paper' and have them play the game.

WHOLE-CLASS INTRODUCTION

FORMAL MEASURES – SQUARE CENTIMETRE

You will need: two copies of BLM 10 'Grid Paper' (one with your hand or foot clearly traced on it, the other blank)

Display a blank copy of BLM 10 'Grid Paper' and ask, 'What do we call each of these little squares? The length of each side is 1 cm.' Discuss the square centimetre and its use in providing a consistent and reproducible measure of area. Use a ruler to draw a 3 × 4 rectangle onto the gridlines. Ask, 'What is the area of this rectangle?' Discuss, then shade and count each square centimetre. Then, present the copy of BLM 10 'Grid Paper' your hand/foot traced on it and ask 'What is the area of my hand/foot?' Invite a student to start shading and counting the whole square centimetres. Once complete, ask, 'What do we do with all of these leftover parts of square centimetres?' Discuss possibilities, and decide what to do with the units that are more than a half and less than a half.

INDEPENDENT TASKS

Note: Choose from Tasks 1, 2 or 3.

You will need: BLM 10 'Grid Paper', NTO 3.23 'Area of Shapes', Student Book p. 93 'Area of Leaves'

TASK 1: AREA OF MY FOOT

Give students BLM 10 'Grid Paper' and have them trace around their foot. Have them estimate the area of their foot in square centimetres. Students colour all of the whole square centimetres in one colour and total the amount. They then look at the remaining squares that are less than and more than a half, and find a method to combine and include these in the total foot area.

TASK 2: INTERACTIVE TASK 

Have students work independently on computers, using NTO 3.23 'Area of Shapes' with the gridlines turned on to experiment with selecting shapes and shading their area. Challenge them to use the shape tools to create a shape that has an area of exactly 16 grid squares.

TASK 3: STUDENT BOOK p. 93 *'Area of Leaves'*

TEACHING GROUP

You will need: BLM 61 'Area of Rectangles', BLM 10 'Grid Paper'

COUNTING SQUARE CENTIMETRES

- For students who require support, have them cut out the rectangles on BLM 59 'Area of Rectangles' and predict their order from biggest to smallest. Have students shade and count the number of square centimetres inside each rectangle, and write this number inside the shape as A = ___ cm^2. Check the order predicted and make changes if necessary.

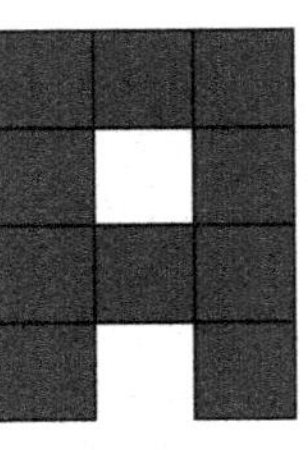

AREA OF LETTERS

- For students who require a challenge, ask them to write their name using the grid paper, where the total area for each letter is 10 cm^2 (see example at right). They may use half squares (divided diagonally or straight) as long as the total equals exactly 10 cm^2.

REFLECTION

Select from the following to suit your class and their learning outcomes:

- Ask, 'What did you do with the square centimetres that weren't fully covered?' and have students share various strategies. Discuss the strengths and limitations of each strategy and decide which would be the most accurate method.
- Ask, 'Do you think it is ever possible to be completely accurate when measuring the area of an irregular shape?'

LESSON PLAN 3

TUNING IN

JAPANESE FLOOR SPACE

You will need: dominoes or rectangles cut from BLM 63 'Tatami Mats' (enough for eight per student)

Say, 'In Japan, floor space is measured in tatami mats. These are rectangular mats that are arranged in different ways to create a square or rectangle. One of the traditional room sizes is 8 mats.' Give each students eight dominoes/paper rectangles each, and have them create a layout of eight mats to create a square or rectangle (see example below).

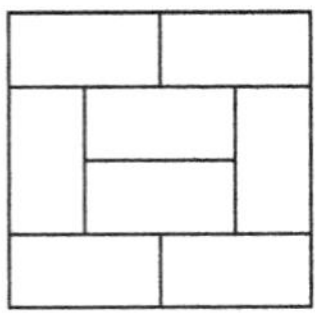

WHOLE-CLASS INTRODUCTION

A SQUARE METRE

You will need: a square centimetre cut from coloured paper, a square metre made from sticking poster paper together

Hold up the square centimetre and ask, 'What do we call this?' The class should recognise this from the previous lesson. Ask, 'Why do you think it is called a square centimetre?' Discuss, emphasising that it is a square, rather than just being a line. Ask, 'If this is a square centimetre, what do you think a square metre would look like?' Discuss. Present the square metre to the class. Discuss the properties of the square metre – each side is exactly 1 m long.

INDEPENDENT TASKS

Note: Choose from Tasks 1, 2 or 3.

You will need: newspapers, sticky tape, LO: *L384 'Finding the area of rectangles'*, Student Book p. 94 'Square Metre – m^2'

TASK 1: MAKING A SQUARE METRE

Have students work in groups of three. Give each group newspaper and sticky tape and ask them to construct a square metre. Then, ask them to estimate how many students can sit down, stand or lie down on a square metre. Have them list items that could be measured with a square metre.

TASK 2: INTERACTIVE TASK 

Have students work independently on computers, using LO: *L384 'Finding the area of rectangles'* to estimate and find the area of rectangles in square centimetres. Note: students will be using a formula to find the area in this activity.

TASK 3: STUDENT BOOK p. 94 ***'Square Metre – m^2'***

TEACHING GROUP

You will need: a constructed square metre, BLM 63 'Tatami Mats', A3 poster paper

BIGGER THAN, SMALLER THAN

- For students who require support, spend more time looking at the square metre. Have students measure each side to ensure they understand its size. Ask students each to find something that is smaller than

1 m^2 and report back to the group. Then have them find an object than is bigger than 1 m^2 and finally, something that is about 1 m^2. Share and discuss all findings with the group.

TATAMI MATS

- For students who require a challenge, have them research the Japanese system of measuring floor space using tatami mats, focusing specifically on traditional layouts. Give students BLM 63 'Tatami Mats' and have them follow the instructions to create floor areas in tatami mats. Have them present their work on an A3 poster to display in the classroom.

REFLECTION

Select from the following to suit your class and their learning outcomes:

- Ask, 'How much bigger is a square metre than a square centimetre?'
- Ask, 'Can you think of any real-life examples where you would use square metres to measure area?'

Home Tasks

Select from the possible Home Tasks:

- Have students find five rectangular items at home that have an area between 50 cm^2 and 80 cm^2. Have them draw the items, write the area of each and list them from biggest to smallest.
- Have students work with their parents or carers to measure the area of their room in square metres.

Assessment

- Have students complete **Student Assessment p. 95**.
- Review with students **Assessment Task Card 3.23**.

During the three lessons:

- Take photos of students undertaking activities to add to their digital portfolios. Students could be invited to write a comment about the activity and their learning.
- Note students' involvement in and ability to complete the interactive activities.
- Review Student Book pages and make notes of areas of difficulty.

Recommendations for Future Learning

Specific to Student Assessment p. 95; if the student is experiencing some difficulty:

Q 1 Have the student examine rulers and tape measures to look at the units marked on them. Remind the student that we are no longer measuring length, but area, so the measure is in 'squares'. Revise content from Lesson Plans 2 and 3.

Q 2 Remind the student of the types of informal measures used in Lesson Plan 1 and the potential problems associated with these.

Q 3–4 Revise activities using the square centimetre from Lesson Plans 1 and 2. Remind the student that area is the size of a surface. Encourage the student to count squares, marking counted squares as they go.

Q 5–6 Revise activities using the square metre from Lesson Plan 3. Show the model of the square metre and have the student use this as a guide to think of objects that would be bigger or smaller.

If the student has not achieved the recommended skills for this unit:

1. See **Assessment Task Card 3.23** for specific recommendations.
2. Have the student work with finding the area of rectangles and squares on grid paper before moving on to irregular shapes.
3. Have the student use a range of measuring devices (including centimetre grid paper) as often as possible to enhance their understanding of the size of each unit.
4. Review *Nelson Maths: Australian Curriculum NSW Year 2* Unit 25.

If the student has achieved the recommended skills and these skills are firmly established, consider:

1. Having the student work with more complex shapes, such as compound shapes.
2. Having the student work with a ruler to find the area of rectangles.
3. Moving forward to *Nelson Maths: Australian Curriculum NSW Year 4* Unit 14.

Division

Number and Algebra
Multiplication and division MA2-6NA uses mental and informal written strategies for multiplication and division

divide, division, equal groups, how many in each group, hundreds, left over, MAB, number of groups, number sentence, ones, remainder, renaming, repeated subtraction, share between, tens, think board, total, worded problem

LESSON PLAN

TUNING IN

WHAT IS DIVISION?

You will need: counters

Record the division symbol on the board. Ask, 'What does this sign mean?' On the board, record responses, e.g. 'shared between', 'divided by', 'how many groups of?' Present the problem: 'I had 14 strawberries and I shared them between two bowls. How many were in each bowl?' Discuss strategies for solving the problem. Have students use counters to model the division situation.

WHOLE-CLASS INTRODUCTION

MODELLING SHARING

You will need: NTO 3.7 'Modelling with MAB'

Present the problem: 36 ÷ 3. Ask, 'How could we model and solve this problem?' Discuss strategies such as sharing 36 counters into three groups. Discuss how to model this division problem using MAB. Have a student use NTO 3.7 'Modelling with MAB' to demonstrate sharing the 3 tens and 6 ones into 3 groups. Repeat with the equations 48 ÷ 4 and 46 ÷ 2. Then present the problem: 52 ÷ 4. Discuss and model how when the tens are shared into four groups there is still one ten remaining. Ask, 'What should we do with this leftover ten?' Model exchanging the ten for 10 ones. Say, 'Now we have 12 ones. How will we share these between four groups?' Repeat with the problems 48 ÷ 3 and 65 ÷ 5.

INDEPENDENT TASKS

Note: Choose from Tasks 1, 2 or 3.

You will need: BLM 64 'Division Problems', MAB or NTO 3.7 'Modelling with MAB', LO: *L2809 'Divide it up: sharing tool'*, Student Book p. 96 'Division Problems'

TASK 1: MODEL IT

Have students take a card made from BLM 64 'Division Problems', then model and solve the division problem using MAB or NTO 3.7 'Modelling with MAB', sharing the total amount into the appropriate number of groups.

TASK 2: INTERACTIVE TASK 

Have students work on computers, using LO: *L2809 'Divide it up: sharing tool'* to explore forming division problems and sharing amounts into equal groups.

TASK 3: STUDENT BOOK p. 96 ***'Division Problems'***

TEACHING GROUP

You will need: BLM 60 'Think Board', NTO 3.7 'Modelling with MAB'

THINK BOARDS

- For students who require support, present the equation: 12 ÷ 3 = 4. Discuss what each number in the equation represents, e.g. 12 is the total number in the collection, 3 is the number of groups and 4 is the number in each group. Have students represent division problems in different ways using BLM 60 'Think Board'. Present them with an equation, e.g. 15 ÷ 3, and have them record the number sentence, a worded problem, a picture and a representation of the materials. Have students explain what each number in their equation represents. Repeat with other equations, e.g. 25 ÷ 5, 18 ÷ 6, 21 ÷ 3.

USING PLACE VALUE

- For students who require a challenge, have them solve division problems with larger numbers. Present the problem: 456 ÷ 4. Use NTO 3.7 'Modelling with MAB' to model sharing the 400 between four groups. Ask, 'Can we share 50 between four groups?' Model sharing a ten to each group and then ask, 'What can we do with the remaining ten?' Have a student model how it can be exchanged for 10 ones. Emphasise how there are now 16 ones that can be shared into the four groups. Repeat with the equation: 462 ÷ 3. Discuss and model how this problem requires renaming in the hundreds and tens. Present the following problems for students to solve using MAB: 575 ÷ 5, 678 ÷ 2, 576 ÷ 4.

REFLECTION

Select from the following to suit your class and their learning outcomes:

- Have students share their work and explain the strategies they used for solving division problems by sharing.
- Have students reflect on Independent Tasks, Task 1, and explain why they had to exchange tens for 10 ones when solving division problems.

LESSON PLAN

TUNING IN

TWO PROBLEMS

You will need: counters

Present the following problems: 'I had 18 apples. I shared them between 6 bags. How many were in each bag?' and 'I had 18 apples. I put 3 apples in each bag. How many bags did I use?' Have students discuss the similarities and differences between the two problems. Ask, 'What information are we provided with in each problem? What are they asking us to find out?' Establish that the first problem tells us how many groups but asks how many in each group, and the second problem tells us how many there are in each group but asks how many groups there are. Solve both problems using counters and discuss what students notice.

WHOLE-CLASS INTRODUCTION

HOW MANY LOTS?

You will need: stickers

Present the worded problem: 'The teacher had 15 stickers. She gave 5 to each student. How many students were there?' Record the equation: 15 ÷ 5. Discuss how this division problem is asking how many lots of 5 there are in 15. Have students discuss strategies for solving the problem. Use stickers to repeatedly form lots of 5 out of the 15 stickers. Discuss how there are three groups of 5 in 15.

INDEPENDENT TASKS

Note: Choose from Tasks 1, 2 or 3.

You will need: counters, BLM 64 'Division Problems', *Paint* or *Kid Pix*, Student Book p. 97 'Building Blocks'

TASK 1: WHAT'S IN 24?

Tell students that 24 is their total number. Record the numbers 6, 4, 12, 2, 8, 3, 24 and 1 on the board and have students work out how many groups of each number can be made from 24, e.g. how many lots of 6 can be made from 24? Encourage students to use repeated subtraction to solve the problems. Have students record the division equations, e.g. 24 ÷ 6 = 4.

TASK 2: INTERACTIVE TASK 

Have students select a card made from BLM 64 'Division Problems', e.g. 20 ÷ 4. Have students model the problem using *Paint* or *Kid Pix*, e.g. using pictures to show how many groups of 4 can be made from 20.

TASK 3: STUDENT BOOK p. 97 ***'Building Blocks'***

TEACHING GROUP

You will need: NTO 3.6 'Number Line', BLM 59 'Multiplication Pictures'

NUMBER LINE

- For students who require support, have them solve division problems using NTO 3.6 'Number Line'. Present the problem: 21 ÷ 7. Have students identify the number 21 on the number line. Establish that the equation is asking how many 7s there are in 21. Ask, 'How many numbers should we jump over each time?' Model skipping over seven numbers from 21 to get to 14, then again to get to 7, and then to zero. Identify that three jumps were made. Ask, 'What is the answer to the problem?' Have students solve the following problems using the number line: 24 ÷ 3, 18 ÷ 9, 16 ÷ 4.

WRITE THE PROBLEMS

- For students who require a challenge, give them BLM 59 'Multiplication Pictures'. Have them write worded division problems to match the pictures. Explain that the worded problems need to include the number in each group and ask how many groups there are, e.g. 'There are 20 kittens. There are 5 kittens in each basket. How many baskets are there?' Have students swap and solve problems.

REFLECTION

Select from the following to suit your class and their learning outcomes:

- Have students share their work from Independent Tasks, Task 1. Discuss the strategies they used to solve the problem. Ask, 'How does division link with repeated subtraction?'
- Have students reflect on the Independent Tasks and the Teaching Group activities. Ask, 'What have you learned about division? What strategies can you use to solve division problems?'

LESSON PLAN 3

TUNING IN

NOT FAIR

You will need: counters

Collect 15 counters and have students pretend that the counters are strawberries. Select two students and explain that you are going to share the strawberries between them. Share them out one at a time so they both have seven counters each. Hold up the remaining counter and ask, 'What should we do with this last one?' Discuss how if the strawberry is given to one person, it wouldn't be fair as they would have unequal amounts. Explain that it is called the remainder. Record the equation on the board and model how to record the answer. Repeat by sharing 17 counters between five students. Ask, 'What is the remainder?' and have students record the equation and the answer.

WHOLE-CLASS INTRODUCTION

REMAINDER RACE

You will need: cards made from BLM 5 'Digit Cards', NTO 3.10 'Card Flip', MAB

Give each student a card from BLM 5 'Digit Cards' and have them stand in a line on one side of the room. Then present a number using NTO 3.10 'Card Flip', selecting 0–99 cards. Have students divide the number shown by the number on their card, using MAB if necessary. If the answer to their division problem has a remainder, the student may take that number of steps, e.g. for 38 ÷ 5, the student would take three steps. Continue by showing another number. The winner is the first student to reach the other side of the room.

INDEPENDENT TASKS

Note: Choose from Tasks 1, 2 or 3.

You will need: BLM 1 '2-Digit Number Cards', 10-sided dice, MAB, LO: *L2812 'Divide it up: kittens'*, Student Book p. 98 'Is There a Remainder?'

TASK 1: GREEN OR RED

Have students turn over a card from BLM 1 '2-Digit Number Cards' and roll a 10-sided dice. Have them form a division problem using the two numbers, e.g. 48 ÷ 5. Students solve and record the division problem, using MAB to model if necessary. Have students circle the problem in green if it has no remainder and red if it has a remainder.

TASK 2: INTERACTIVE TASK

Have students work on computers, using LO: *L2812 'Divide it up: kittens'* to explore solving division problems by sharing into equal groups and identifying remainders.

TASK 3: STUDENT BOOK p. 98 *'Is There a Remainder?'*

TEACHING GROUP

You will need: paper plates, small kinder squares, deck of playing cards (picture, 10 and joker cards removed), butcher's paper

SANDWICHES ON PLATES

- For students who require support, give them paper plates and small kinder squares. Have them pretend the kinder squares are sandwiches. Explain that they need to share the sandwiches evenly between the plates. Present the problem: 25 ÷ 4. Have students collect and share 25 sandwiches between four plates. Ask, 'How many sandwiches are left over?' Have students solve the following problems: 16 ÷ 3, 25 ÷ 2, 27 ÷ 4, 32 ÷ 7, 19 ÷ 4.

MAKING REMAINDERS

- For students who require a challenge, write the headings 'Remainder 1', 'Remainder 2', 'Remainder 3' and so on up to 'Remainder 9' each on a sheet of butcher's paper. Present students with a deck of playing cards (picture, 10 and joker cards removed) and have them select three cards to make a 2-digit and 1-digit number, e.g. 53 and 5. Have them form and solve division problems and then record them on the appropriate sheet of paper, depending on the remainder. Ask, 'How many problems can you make for each category?'

REFLECTION

Select from the following to suit your class and their learning outcomes:

- Present the equation: 21 ÷ 4. Have students solve the problem. Ask, 'What is the remainder? What strategies did you use to solve the problem?' Have them explain why a division problem may have a remainder.
- Have students share their work from the Teaching Group activity 'Making Remainders'. Ask, 'What strategies did you use to identify division problems with remainders?'

Home Tasks

Select from the possible Home Tasks:

- Have students identify and solve real-life division situations, e.g. 'We have 10 lollies. How could we share them fairly between 5 children?'
- Have students find 2-digit numbers in catalogues and magazines. Encourage them to divide the numbers by 4 and identify the problems with remainders.

Assessment

- Have students complete **Student Assessment p. 99**.
- Review with students **Assessment Task Card 3.24**.

During the three lessons:

- Observe which students are able to solve division problems by forming groups or sharing into groups. Identify which students are able to solve division problems with remainders.
- Make note of students who completed the scaffolding tasks or the more challenging activities of the Teaching Groups.
- Review Student Book pages and make notes of areas of difficulty.

Recommendations for Future Learning

Specific to Student Assessment p. 99; if the student is experiencing some difficulty:

Q 1–2 Revise the strategy of solving division problems by sharing the total amount into the appropriate number of groups. Encourage the student to use materials such as counters or MAB to model the sharing process. Discuss how the answer is the number of items in each group.

Q 3–4 Revise the strategy of solving division problems by identifying the number of equal groups that can be made out of a number. Discuss how some division problems require us to find out the number of groups, rather than the number in each group. Use a number line to model.

Q 5 Review how there are sometimes leftovers when solving division problems. Emphasise how the groups need to be equal. Model division problems using materials and have the student identify remainders.

If the student has not achieved the recommended skills for this unit:

1. See **Assessment Task Card 3.24** for specific recommendations.
2. Have the student use materials such as MAB or counters to model division problems.
3. Review *Nelson Maths: Australian Curriculum NSW Year 2* Unit 27.
4. Have the student complete *Building Mental Strategies Skill Book Year 3*, pp. 68–71, to reinforce mental strategies for division.

If the student has achieved the recommended skills and these skills are firmly established, consider:

1. Having the student complete *Building Mental Strategies Skill Book Year 4*, pp. 66–67, to further explore mental strategies for division.
2. Moving forward to *Nelson Maths: Australian Curriculum NSW Year 4* Unit 15.
3. Extending the student by introducing division involving larger numbers.

Unit 25 More About Division

Number and Algebra
Multiplication and division MA2-6NA uses mental and informal written strategies for multiplication and division

array, divide, division, equation, fact, fact family, groups, multiplication, multiply, number line, skip counting

LESSON PLAN 1

TUNING IN

QUICKEST ARRAY

You will need: sticky notes, counters

Give each student a sticky note. Explain to them that 10 is the largest number they may use. Tell half the class to write a number of rows on their sticky note, e.g. 4 rows. Tell the other half to write how many in each row on their sticky note, e.g. 4 in each row. Have students move around the room and when you say 'array', they must find a partner, i.e. a student with a number of rows on their sticky note would go with a student with how many in each row, and solve their array. Give students counters if necessary. The first pair to solve their array correctly wins a point. Continue with students finding a different partner each time.

WHOLE-CLASS INTRODUCTION

SOLVING WITH ARRAYS

You will need: NTO 3.20 'Counters'

Use NTO 3.20 'Counters' to present students with an array of five rows of four counters. Write the equation: 20 ÷ 5. Ask, 'How can we use the array to solve this problem?' Discuss how the equation tells us that there are 20 altogether and there are five rows, so that means there are four in each row. Discuss how arrays can be used to represent division problems. Present the problem: 16 ÷ 4. Ask, 'How could we use an array to help us solve this problem?' Test and discuss students' ideas.

INDEPENDENT TASKS

Note: Choose from Tasks 1, 2 or 3.

You will need: BLM 65 'Array Cards', calculators, NTO 3.20 'Counters', Student Book p. 100 'Arrays for Division'

TASK 1: ARRAY RACE

Give pairs of students cards made from BLM 65 'Array Cards' and place them face down. They turn over a card, record a division problem to match the array, then use a calculator to check their answers. The first student to form the equation correctly wins a point. Continue with other cards.

TASK 2: INTERACTIVE TASK 

Have students form different arrays using NTO 3.20 'Counters' and record the division and multiplication problems that would match each array.

TASK 3: STUDENT BOOK p. 100 *'Arrays for Division'*

TEACHING GROUP

You will need: dice, BLM 10 'Grid Paper', BLM 64 'Division Problems'

CUT OUT ARRAYS

- For students who require support, have them roll two dice and use the numbers shown to form an array, e.g. six rows of 4. Have students cut out their array from BLM 10 'Grid Paper', then record a multiplication problem and a division problem to match their array.

SOLVE IT WITH ARRAYS

- For students who require a challenge, give them BLM 10 'Grid Paper' and cards made from BLM 64 'Division Problems'. Ask, 'Could we make arrays to help us solve these division problems?' Discuss students' strategies and ideas. Have them select cards and draw arrays on the grid paper to solve the division problems.

REFLECTION

Select from the following to suit your class and their learning outcomes:

- Present students with a 4 × 3 array. Have them explain how the array can relate to multiplication and division problems. Ask, 'How can an array help you solve division problems?'
- Have students reflect on Independent Tasks, Task 2. Ask, 'What did you notice about each array, multiplication problem and division problem that you recorded?'

LESSON PLAN 2

TUNING IN

USING MULTIPLICATION

Present the problem: 45 ÷ 5. Ask, 'How could we use multiplication to solve this division problem?' Discuss students' ideas and emphasise the connection between multiplication and division by drawing an array of 9 × 5. Discuss how they could use the familiar fact 9 × 5 = 45 to solve the problem. Repeat with the problem 24 ÷ 3.

WHOLE-CLASS INTRODUCTION

FEET OR KNEES

You will need: counters

Explain to students that they are going to be shown multiplication and division problems with the operation symbol missing. They need to stand up if they think it is a multiplication problem and kneel down if they think it is a division problem. Present the problem: 24 ___ 6 = 4. Discuss why students decided to stand or kneel. Use counters to show the answer. Repeat with other problems, e.g. 5 ___ 6 = 30, 16 ___ 4 = 4, 3 ___ 7 = 21.

INDEPENDENT TASKS

Note: Choose from Tasks 1, 2 or 3.

You will need: BLM 66 'Multiply or Divide', 10-sided dice, BLM 51 'Multiplication Fact Cards', NTO 3.11 'Calculator', Student Book p. 101 'Division Riddles'

TASK 1: MULTIPLY OR DIVIDE

Give pairs of students BLM 66 'Multiply or Divide' and two 10-sided dice. Students take turns to roll the two dice. They can choose to multiply or divide the two numbers rolled and then record the equation in their box. If the answer to their problem is on the number mat, then they may cross it off. If not, it is their partner's turn. The winner is the student who crosses off the last number on the number mat.

TASK 2: INTERACTIVE TASK 

Give students cards made from BLM 51 'Multiplication Fact Cards' and place them face down. They turn over a card, solve the multiplication fact and then record the division facts which that multiplication fact would help them solve. Have them use NTO 3.11 'Calculator' to check if the division problem is correct.

TASK 3: STUDENT BOOK p. 101 *'Division Riddles'*

TEACHING GROUP

You will need: NTO 3.6 'Number Line'

FORWARDS AND BACKWARDS

- For students who require support, present the equations: 20 ÷ 5, 4 × 5. Model solving the division problem by skip counting back by 5s using NTO 3.6 'Number Line'. Then model solving the multiplication problem by counting forwards by 5s to get to 20. Discuss what students notice. Have students use the number line to solve related problems, e.g. 27 ÷ 9, 9 × 3.

HOW MANY PROBLEMS?

- For students who require a challenge, present them with the number 24. Explain that they need to find all the division problems that involve the number 24, e.g. 24 ÷ 2 = 12, 24 ÷ 3 = 8. Ask, 'How could you use multiplication to help you solve this problem?' Repeat with the numbers 32, 20 and 48.

REFLECTION

Select from the following to suit your class and their learning outcomes:

- Have students explain how they can use multiplication to help solve division problems.
- Have students reflect on Independent Tasks, Task 1. Discuss strategies they used for deciding whether to multiply or divide the numbers on the dice.

TUNING IN

WHAT DO YOU NOTICE?

You will need: counters

Write the following equations on the board: 12 ÷ 3 = 4, 12 ÷ 4 = 3, 4 × 3 = 12, 3 × 4 = 12. Give students 12 counters and have them model the four problems. Ask, 'What do you notice about these four equations?' Discuss how these equations form a multiplication and division fact family because they use the same numbers. Have students discuss the relationship between the equations. Present the numbers: 5, 2, 10. Ask, 'What fact family could you make with these numbers?'

WHOLE-CLASS INTRODUCTION

STEP IT OUT

You will need: masking tape or chalk

Mark a large 4 × 4 grid on the floor using masking tape or chalk. Write the numbers from zero to 12 and division, multiplication and equals symbols in the squares, so that each square has either a number or a symbol. Choose a student to step out a division number sentence on the grid, e.g. 8 ÷ 4 = 2. Select another student to step out a related multiplication or division fact. Continue so that all four related multiplication and division facts are covered. Discuss what students notice. Repeat with a student stepping out a different equation.

INDEPENDENT TASKS

Note: Choose from Tasks 1, 2 or 3.

You will need: BLM 5 'Digit Cards', blank squares of paper, NTO 3.17 'Ten-Sided Dice', Student Book p. 102 'Puzzle Pieces'

TASK 1: REARRANGE

Give students cards made from BLM 5 'Digit Cards' and blank squares of paper. Have them draw a multiplication symbol, a division symbol and an equals symbol each on a square of paper. Students form and record a multiplication or division fact using the cards and the paper squares, e.g. 5 × 3 = 15. Have them then rearrange the cards and paper squares so they form and record the four related facts, e.g. 3 × 5 = 15, 15 ÷ 3 = 5 and 15 ÷ 5 = 3. Have students select three coloured pencils and colour matching numbers in the equations the same colour, e.g. blue for 5s, yellow for 3s and red for 15s.

TASK 2: INTERACTIVE TASK 

Give students cards made from BLM 5 'Digit Cards' and place them face down. Students turn over a card and show a number using NTO 3.17 'Ten-Sided Dice'. Have them multiply these numbers together and record the equation, then record the related multiplication and division facts.

TASK 3: STUDENT BOOK p. 102 ***'Puzzle Pieces'***

TEACHING GROUP

You will need: counters, BLM 17 'Blank Dominoes', BLM 58 'Multiplication Cards'

FOUR EQUATIONS

- For students who require support, present them with the numbers: 6, 3, 18. Explain that there are four equations that can be made using these three numbers. Have students use counters to model the two multiplication and two division problems. Ask, 'How do we know that these equations are related?' Have students record the problems for other fact family numbers, e.g. 7, 3, 21; 4, 9, 36.

FACT FAMILY DOMINOES

- For students who require a challenge, give them BLM 17 'Blank Dominoes' and cards made from BLM 58 'Multiplication Cards', placed face down. Have them turn over a card and copy the equation onto one half of a blank domino. Have students record a related division equation on half of a different blank domino. Tell students that they do not need to record the answers on the dominoes. When all dominoes are completed, students cut them out and join with a partner. They shuffle the dominoes, deal five to each player and place the remaining dominoes in a pile. Students then take turns to join the dominoes by matching a multiplication problem with the related division problem. If they are unable to place a domino, they pick one up from the pile. The winner is the first person to join all their dominoes.

REFLECTION

Select from the following to suit your class and their learning outcomes:

- Have students reflect on Independent Tasks, Task 1. Ask, 'What did you notice about the numbers you were colouring in the equations? How do you know the problems are related?'
- Have students explain how multiplication problems can help solve division problems. Ask, 'If you know the fact 4 × 5 = 20, what other problems would you be able to solve?'

Home Tasks

Select from the possible Home Tasks:

- Have students find examples of multiplication and division problems around the home, e.g. there are eight pillows shared between four beds. Have them write both the multiplication and division equations to match the situation.
- Have students find numbers around the home and form division problems, but use a related multiplication fact to solve the problem.

Assessment

- Have students complete **Student Assessment p. 103**.
- Review with students **Assessment Task Card 3.25**.

During the three lessons:

- Observe which students are able to form multiplication and division fact families. Identify students who can use arrays to solve division problems.
- Make note of students who completed the scaffolding tasks or the more challenging activities of the Teaching Groups.
- Review Student Book pages and make notes of areas of difficulty.

Recommendations for Future Learning

Specific to Student Assessment p. 103; if the student experiencing some difficulty:

Q 1–2 Review how arrays relate to division situations. Give the student some arrays and have them identify the total amount, the number of rows and the number in each row. Have the student write the division equations. Revise how an array can be used to solve a division problem.

Q 3 Revise the connection between division and multiplication. Discuss how a familiar multiplication fact can help solve a division fact if they are related.

Q 4–5 Revise multiplication and division fact families. Have the student use counters to model the multiplication and division problems and identify how the equations are similar. Emphasise how the problems are related because they involve the same numbers.

If the student has not achieved the recommended skills for this unit:

1. See **Assessment Task Card 3.25** for specific recommendations.
2. Have the student use materials such as MAB or counters to model multiplication and division problems and to identify their connection.
3. Review *Nelson Maths: Australian Curriculum NSW Year 2* Unit 27.
4. Have the student complete *Building Mental Strategies Skill Book Year 3*, pp. 70–71, to reinforce using multiplication facts to solve division.

If the student has achieved the recommended skills and these skills are firmly established, consider:

1. Having the student complete *Building Mental Strategies Skill Book Year 4*, pp. 66–68, to further explore using multiplication facts to solve division.
2. Moving forward to *Nelson Maths: Australian Curriculum NSW Year 4* Unit 11.

Unit 26 Fractions

Number and Algebra
Fractions and decimals MA2-7NA represents, models and compares commonly used fractions and decimals

divide into, partition, equal parts, eighth, fifth, half, quarter, third and other fractions at your discretion

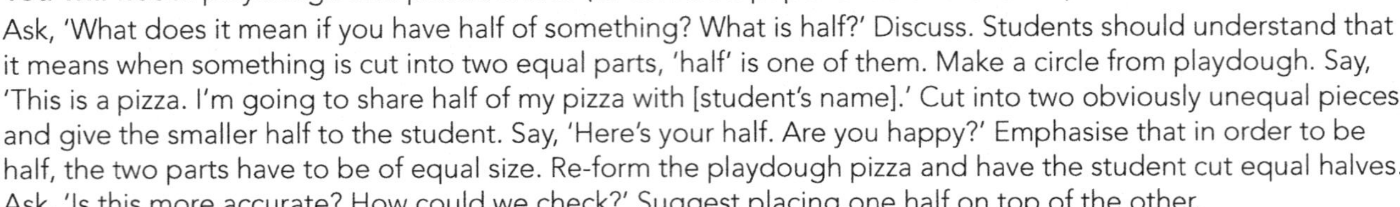

LESSON PLAN 1

TUNING IN

UNDERSTANDING HALF

You will need: playdough and plastic knives (or coloured paper circle and scissors)

Ask, 'What does it mean if you have half of something? What is half?' Discuss. Students should understand that it means when something is cut into two equal parts, 'half' is one of them. Make a circle from playdough. Say, 'This is a pizza. I'm going to share half of my pizza with [student's name].' Cut into two obviously unequal pieces and give the smaller half to the student. Say, 'Here's your half. Are you happy?' Emphasise that in order to be half, the two parts have to be of equal size. Re-form the playdough pizza and have the student cut equal halves. Ask, 'Is this more accurate? How could we check?' Suggest placing one half on top of the other.

WHOLE-CLASS INTRODUCTION

MAKING EQUAL PARTS

You will need: playdough and plastic knives (or coloured paper shapes and scissors)

Write $\frac{1}{2}$ on the board. Ask, 'What other fractions do you know?' List unit fractions on the board, focusing on $\frac{1}{2}$, $\frac{1}{4}$, $\frac{1}{3}$, $\frac{1}{5}$ and $\frac{1}{8}$, and name them 'one third, one quarter' etc. Make a square out of playdough and ask, 'How could I cut this into quarters? How many equal parts would I need?' Select one student to demonstrate. Examine the quarters with the class and ask, 'Is a quarter bigger or smaller than a half?' Discuss. Ask, 'Is there another way we could cut the square into quarters?' Invite students to demonstrate other methods. Continue making a range of shapes and having students partition them into thirds, quarters and fifths and compare their sizes. Emphasise that the more equal parts something is cut into, the smaller each part will be.

INDEPENDENT TASKS

Note: Choose from Tasks 1, 2 or 3.

You will need: coloured A5 paper and kinder squares or circles, scissors, A3 paper, *Paint,* Student Book p. 104 'Drawing Fractions'

TASK 1: FOLDING FRACTIONS

Using square, rectangular or circular coloured paper, have students fold and cut five different fractions from the list displayed on the board ($\frac{1}{2}$, $\frac{1}{4}$, $\frac{1}{3}$, $\frac{1}{5}$ and $\frac{1}{8}$). Emphasise the need for equal parts. Have them paste their partitioned shapes onto A3 paper, then shade and label the fraction, e.g. $\frac{1}{3}$. Students add three fraction facts based on their work to the poster, e.g. $\frac{1}{3}$ is bigger than $\frac{1}{4}$.

TASK 2: INTERACTIVE TASK 

Have students work independently on computers, using *Paint* to make as many representations of 'half' as they can. Students should use the 'Shape' function to draw shapes and straight lines to divide them into equal parts, then use 'Fill' to shade half of each shape.

TASK 3: STUDENT BOOK p. 104 *'Drawing Fractions'*

TEACHING GROUP

You will need: playdough, plastic knives, coloured A5 paper and kinder squares or circles, scissors, A3 paper

PLAYDOUGH FRACTIONS

- For students who require support, make five same-size playdough squares. Ask, 'If I want to cut my squares in half, what do I need to do?' Discuss with students. Cut the squares, equally and unequally:

Ask, 'Have I cut these in half?' Emphasise the need for equal parts. Have students identify those that are cut in half. Students make a playdough square of their own and partition it in half, them divide each half equally into half again. Ask, 'How many parts do we have now?' Discuss the partition of a shape into four parts being quarters. Have students model other partitions with the dough.

PAPER FOLDING CHALLENGE

- For students who require a challenge, have them create more challenging folded fractions – sixths, sevenths, ninths and tenths – modelling each fraction using two different paper shapes. Have them cut out and paste their partitioned shapes onto an A3 poster, and shade to show both unit, e.g. $\frac{1}{7}$, and non-unit, e.g. $\frac{4}{4}$, fractions. Have them add three fraction facts based on their work to their poster.

REFLECTION

Select from the following to suit your class and their learning outcomes:

- On the IWB, open *Paint* and use the 'Shapes' tool to draw a hexagon. Ask, 'How many different ways could we divide this shape in half?' Select and copy the shape five times and use the line tool to show the different ways to divide the shape in half (there are six ways).
- Ask, 'What does the bottom number (the denominator) on a fraction tell us?' Emphasise again the partitioning into equal parts.

LESSON PLAN 2

TUNING IN

HOW MANY PARTS?

Draw a rectangle on the board, divide it into five equal parts then shade four of them. Ask, 'What fraction of the shape has been shaded?' Correct answers are: 'four fifths', 'or four out of five'. Extend students' thinking by asking, 'How many more parts would you need to make a whole?' Then ask, 'Can we show a half of the rectangle using fifths?' Discuss. Repeat this process using different shapes, e.g. a circle divided into eighths, a square into quarters, with different numbers of parts shaded.

WHOLE-CLASS INTRODUCTION

A NUMBER OF EQUAL PARTS

You will need: NTO 3.24 'Fraction Circles'

Display NTO 3.24 'Fraction Circles' and show the fraction $\frac{1}{4}$. Select one student to name the fraction. Ask, 'What does the bottom number represent?' Students should know the bottom number (the denominator) tells us how many equal parts the number is divided into. Then ask, 'What does the top number tell us?' To aid thinking, change the fraction to $\frac{2}{4}$ and watch the model change. Explain that the top number (the numerator) tells us how many equal parts we are talking about. Repeat by changing the fraction, using different denominators and having students predict what the model will look like with various numerators. Finish by covering the fraction made and having students work out what fraction is shown.

INDEPENDENT TASKS

Note: Choose from Tasks 1, 2 or 3.

You will need: BLM 67 'Fraction Wall', 10-sided dice, coloured pencils, LO: *L2801 'Fraction fiddle: matching cake fractions'*, Student Book p. 105 'Fractions 4 Ways'

TASK 1: FRACTION WALL GAME

Have students work in pairs with BLM 67 'Fraction Wall', two 10-sided dice and two different coloured pencils (one for each student). The students take turns to roll the dice and create a fraction (bigger number = denominator, smaller number = numerator), then find and colour this fraction on the fraction wall. They are only allowed to colour their fraction if there is space on the board for it – no overlapping is allowed. If a fraction doesn't fit, they miss their turn. Once the board is full, players count up the total number of parts they shaded; the highest score wins.

TASK 2: INTERACTIVE TASK 

Have students work independently on computers, using LO: *L2801 'Fraction fiddle: matching cake fractions'* to predict and check the fraction of cake displayed in the diagram.

TASK 3: STUDENT BOOK p. 105 *'Fractions 4 Ways'*

TEACHING GROUP

You will need: Unifix blocks, BLM 67 'Fraction Wall', 10-sided dice, coloured pencils

FRACTION BLOCKS

- For students who require support, model $\frac{1}{4}$ with Unifix blocks by joining three red blocks and one yellow block together. Ask, 'How many blocks are there altogether? Four. So my whole shape has four equal parts. How many of the blocks are yellow? What fraction is yellow?' Discuss. Ask, 'How do we write this fraction?' Reinforce the meanings of the bottom and top numbers. Repeat with another unit fraction, e.g. $\frac{1}{5}$. Have students model various unit fractions. If/when ready, move on to non-unit fractions, e.g. $\frac{2}{3}$.

SIZE COUNTS

- For students who require a challenge, have them work in pairs to play the fraction wall game from Independent Tasks, Task 1, modifying the rule so that the winner is not the player with the most parts, but the player who has shaded the largest value of parts. Students may discuss methods for working out the combined values, but an estimate will suffice.

REFLECTION

Select from the following to suit your class and their learning outcomes:

- Ask, 'What does the top number (the numerator) on a fraction tell us?' Emphasise again that it tells us the number of equal parts we have.
- Ask, 'Which is bigger, $\frac{1}{5}$ or $\frac{1}{2}$? How do you know?'

LESSON PLAN 3

TUNING IN

COMPARING TO HALF

Write the following fractions on the board: $\frac{3}{4}, \frac{1}{5}, \frac{3}{7}, \frac{8}{9}, \frac{4}{5}, \frac{52}{100}, \frac{3}{50}, \frac{1}{7}, \frac{6}{7}, \frac{6}{5}, \frac{2}{3}, \frac{1}{4}$. Have students draw a line down the middle of their page, write 'smaller than half' on one side and 'bigger than half' on the other side and have them write the fractions on the appropriate side. Share results as a class, discussing any tricky numbers and drawing diagrams to help understanding if required.

WHOLE-CLASS INTRODUCTION

EXAMINING FRACTION SIZE

You will need: egg carton, golf or ping pong balls

Show the egg carton to the class and ask, 'How many parts is this carton divided into? Are they all of equal size?' Students should recognise that there are 12 equal parts. Ask, 'How many parts need to be filled to show one whole? One half?' Place eight golf or ping pong balls in the carton, and ask, 'What fraction is shown?' Discuss. Ask, 'Is this fraction smaller than half, equal to half or bigger than half?' Have students respond and provide reasoning. Repeat with other numbers of balls in the carton, e.g. 1, 3, 10.

INDEPENDENT TASKS

Note: Choose from Tasks 1, 2 or 3.

You will need: BLM 68 'Fraction Cards', A3 paper, LO: *L135 'Shape fractions'*, Student Book p. 106 'Fraction Shapes'

TASK 1: EQUAL, SMALLER OR BIGGER THAN HALF?

Give pairs of students cards made BLM 68 'Fraction Cards'. Have them divide the cards into three categories – smaller than half, equal to half and bigger than half – and display these on A3 paper. Students select two cards from each category and write a sentence about how they decided on the category, e.g. 'We knew $\frac{4}{6}$ is bigger than half because $\frac{3}{6}$ is equal to half.'

TASK 2: INTERACTIVE TASK 

Have students work independently on computers, using LO: *L135 'Shape fractions'* to explore partitioning different shapes to make fractions by cutting shapes into a specified number of equal parts, then shading the number required. This also looks at equivalent fractions.

TASK 3: STUDENT BOOK p. 106 *'Fraction Shapes'*

TEACHING GROUP

You will need: BLM 10 'Grid Paper', BLM 68 'Fraction Cards', A3 paper

GRID PAPER SHAPES

- For students who require support, draw a 2 × 2 square on BLM 10 'Grid Paper'. Ask, 'How many equal parts are in this square? How many of these parts would I shade to show $\frac{1}{4}$?' Emphasise that one quarter is one part out of four. Ask, 'What would it look like if we shaded half of the square? Is this bigger or smaller than the quarter? How many more parts would we shade to have one whole?' Have students draw different sized squares and rectangles in their books, shade parts to make fractions and decide whether they are bigger or smaller than half.

FRACTION VALUES

- For students who require a challenge, have them order the cards on BLM 68 'Fraction Cards' from smallest to largest and paste them on a sheet of A3 paper. They may like to use grid paper to draw models of the fractions in order to compare size.

REFLECTION

Select from the following to suit your class and their learning outcomes:

- Ask, 'How did you work out whether fractions were equal, smaller or bigger than half?'
- Ask, 'When is it difficult to divide a shape into equal parts?'

Home Tasks

Select from the possible Home Tasks:

- Have students make a list of times when their family uses fractions in everyday life, e.g. if they have pizza, note how many slices it is divided into; or if treats are shared out, note how many pieces each person receives.
- Have students find examples of fractions from newspapers or magazines and decide whether they are smaller than half, equal to half or bigger than half. They can add these to their A3 posters.

Assessment

- Have students complete **Student Assessment p. 107**.
- Review with students **Assessment Task Card 3.26**.

During the three lessons:

- Take photographs of students manipulating concrete materials (e.g. playdough, paper shapes, blocks) to include in digital portfolios as evidence of partitioning shapes to form fractions.
- Collect the posters from Lesson Plan 3 with the added numbers from the Home Task to demonstrate the learning link between home and school.
- Have students write reflections on how the interactive learning objects helped them understand concepts, to show the use and effectiveness of ICT in the classroom.

Recommendations for Future Learning

Specific to Student Assessment p. 107; if the student is experiencing some difficulty:

Q 1 Review the activity of cutting playdough into two equal parts to form halves. Emphasise the need for the two parts to be of equal size.

Q 2–3 Revisit the Lesson Plan 1 activities from the Whole-Class Introduction and 'Playdough Fractions' from the Teaching Group, using playdough to model both $\frac{1}{2}$ and $\frac{1}{4}$ and compare their size.

Q 4–6 Review Lesson Plans 1 and 2 to gain an understanding of the denominator and numerator. Select a unit fraction and concrete materials to model both.

Q 7 Revisit the Whole-Class Introduction activity in Lesson Plan 3. Have the student draw and shade the fractions on grid paper to decide whether they are close to half, folding the grid paper shape in half to aid thinking.

If the student has not achieved the recommended skills for this unit:

1. See **Assessment Task Card 3.26** for specific recommendations.
2. Have the student use concrete materials to develop their understanding of $\frac{1}{2}$ and $\frac{1}{4}$ before moving on to other unit and non-unit fractions.
3. Review *Nelson Maths: Australian Curriculum NSW Year 2* Unit 7.

If the student has achieved the recommended skills and these skills are firmly established, consider:

1. Having the student complete *Building Mental Strategies Skill Book Year 3*, pp. 6–7.
2. Moving forward to LO: *L2803 'Fraction fiddle: comparing non-unit fractions'*.
3. Moving forward to *Nelson Maths: Australian Curriculum NSW Year 4* Unit 23.

Unit 27 More About Fractions

Number and Algebra
Fractions and decimals MA2-7NA represents, models and compares commonly used fractions and decimals

eighth, fifth, half, ninth, quarter, seventh, sixth, tenth, third, whole

LESSON PLAN 1

TUNING IN

NEWSPAPER FRACTIONS

You will need: a newspaper page with a mix of advertisements, photos and text; three cards; magnets

Hold up the newspaper page and ask students to estimate what fraction of the page is text. Write the fraction on a card. Repeat for photos and advertisements. Use magnets to attach the three cards to the board. Ask, 'How could I draw these fractions? What shape might I use?' Have students volunteer answers and select students to draw the fractions. Ask, 'Do all of these fractions add up to one whole, or thereabouts?' It may be difficult to give an accurate answer if different denominators have been used, so an approximation is fine.

WHOLE-CLASS INTRODUCTION

ESTIMATING FRACTIONS TO CREATE A FRACTION WALL

You will need: six 1 m lengths of thick paper streamers (different colours if possible), coloured cards (plus the three from Tuning In), magnets

Use a magnet to attach a streamer to the board and write '0' and '1' at either end. Write '$\frac{1}{2}$' on one card. Ask, 'Can you estimate where half of the streamer is?' Select a student to attach the card to the correct place. Ask, 'How many halves do you need to make one whole?' Look at the streamer, showing there are two equal halves. Select another streamer and a fraction from the Tuning In activity, e.g. $\frac{1}{3}$. Estimate its location on the streamer. Discuss how many of these are needed to make a whole. Ask, 'What other fractions between zero and 1 have this denominator?' Have students list fractions, e.g. $\frac{2}{3}$, write them on cards and estimate their locations on the streamer. Ask, 'How could we check their accuracy?' Students may suggest measuring or folding the streamer to check for equal parts. Repeat to cover quarters, fifths and sixths. Leave one streamer blank to demonstrate one whole. Display the streamers as a fraction wall in the classroom.

INDEPENDENT TASKS

Note: Choose from Tasks 1, 2 or 3.

You will need: BLM 69 'Making a Fraction Wall', A3 paper, BLM 70 'Digital Fraction Wall', *Word,* Student Book p. 108 'Fraction Wall'

TASK 1: MAKING A FRACTION WALL

Give students BLM 69 'Making a Fraction Wall' and have them follow the instructions to make a fraction wall, then stick it onto A3 paper. On the back, have students write ten statements about fractions, based on the wall – four 'equal to', three 'bigger than' and three 'smaller than' statements, e.g. $\frac{6}{6}$ is equal to one whole; $\frac{8}{10}$ is bigger than $\frac{2}{3}$.

TASK 2: INTERACTIVE TASK 

Give students BLM 70 'Digital Fraction Wall' and have them work independently on computers to make a fraction wall using *Word.*

TASK 3: STUDENT BOOK p. 108 ***'Fraction Wall'***

TEACHING GROUP

You will need: BLM 67 'Fraction Wall', coloured pencils, streamers, cards, paperclips

UNDERSTANDING THE FRACTION WALL

- For students who require support, give them BLM 67 'Fraction Wall'. Have them shade the top row, and ask, 'How much of the top line have you shaded?' (one whole) Have them shade one section of the next row, and ask, 'How much have you shaded?' If students answer 'One', ask, 'One out of how many?' Emphasise that one out of two is one half, written as $\frac{1}{2}$. Ask, 'How many halves equal one whole?' Continue looking at other fractional parts, asking how many equal one whole and comparing equivalent fractions as they arise.

MORE FRACTION STREAMERS

- For students who require a challenge, give them streamers and cards to create fraction strips for sevenths, eighths, ninths and tenths. Have students work together to work out where each fractional part will be located, create fraction cards and attach them to the streamers with paperclips. These can be added to the class fraction wall.

REFLECTION

Select from the following to suit your class and their learning outcomes:

- Ask, 'How many fifths do you need to make a whole?' Repeat for other fractions, e.g. thirds, quarters.
- Have students share their work from the Independent Tasks, and ask 'What did you learn about fractions from the fraction wall? How can you use it to compare sizes of fractions?'

LESSON PLAN 2

TUNING IN

ORDERING UNIT FRACTIONS

Write the following unit fractions on the board: $\frac{1}{6}, \frac{1}{2}, \frac{1}{10}, \frac{1}{4}, \frac{1}{5}, \frac{1}{8}, \frac{1}{3}, \frac{1}{7}, \frac{1}{9}$. Ask students to order the fractions from biggest to smallest. Students may use the class fraction wall to help. Ask, 'Which fraction is the largest? Which is the smallest?' Students should now understand that the more parts a whole is divided into, the smaller each part will be.

WHOLE-CLASS INTRODUCTION

FRACTIONS ON NUMBER LINES

You will need: NTO 3.6 'Number Line'

Display NTO 3.6 'Number Line' with an open number line selected. Mark the ends '0' and '2'. Ask, 'Where is $\frac{1}{2}$ on this number line?' Have students volunteer responses. They may choose the middle of the line rather than the actual value of $\frac{1}{2}$; if so, point out the end values, and ask, 'Where would 1 be? Then, where is $\frac{1}{2}$?' Discuss. Then ask, 'What number is halfway between 1 and 2?' Discuss, and add $1\frac{1}{2}$ to the number line. Next, draw a marker halfway between zero and $\frac{1}{2}$ and ask, 'What fraction is this? It is exactly half as big as $\frac{1}{2}$.' Refer students to the fraction wall for assistance. Discuss and add $\frac{1}{4}$ to the number line.

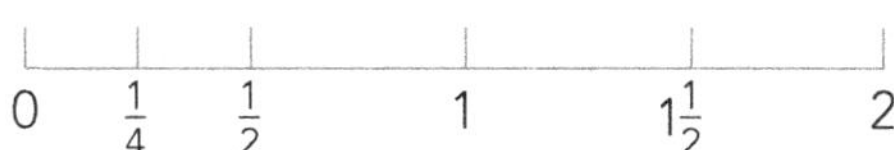

Use the number line to count by quarters from zero to 2 and fill in the gaps. Ask, 'What would come next?' How could I draw a picture showing $1\frac{1}{2}$ on the board?' Select one student to draw the picture on the board, e.g. a whole pizza and half of another pizza. Open a new number line, marking the ends '0' and '3'. Ask, 'Where would $\frac{1}{3}$ be?' Have students locate $\frac{1}{3}$ and $\frac{2}{3}$, then count by thirds to fill the number line.

INDEPENDENT TASKS

Note: Choose from Tasks 1, 2 or 3.

You will need: paper streamers, BLM 71 'Horse Race', LO: *L2802 'Fraction fiddle: comparing unit fractions'*, Student Book p. 109 'Counting by Fractions'

TASK 1: FRACTION RACE

Have students work in groups of four. Give each group a 3 m paper streamer, have them fold it and mark the quarters up to 3, e.g. $\frac{1}{4}, \frac{1}{2}, \frac{3}{4}$, 1, $1\frac{1}{4}$ etc. Use cards from BLM 71 'Horse Race' and colour a horse each. Students select a card and move forwards or back along the track to match the instruction on the card. The winner is the first student to reach 3.

TASK 2: INTERACTIVE TASK 

Have students work independently on computers, using LO: *L2802 'Fraction fiddle: comparing unit fractions'* to compare fractions, e.g. deciding whether one third is larger than one quarter. They then build and compare fractions on a number line.

TASK 3: STUDENT BOOK p. 109 ***'Counting by Fractions'***

TEACHING GROUP

You will need: NTO 3.6 'Number Line'; fraction cards labelled with eighths; *Building Mental Strategies Skill Book Year 3* 'Locating Fractions on a Number Line', pp. 18–19

ZERO TO 1 NUMBER LINES

- For students who require support, display NTO 3.6 'Number Line' and select a number line from 0–1, divided into eight sections. Have students count the sections. Ask, 'How many parts has this line been divided into?' If required, draw a rectangle using the number line as the base, and divide it into eight parts using the markers. Ask, 'What would we name one of these parts?' (one eighth) Place 'eighths' fraction cards face down in a pile, have students select cards and locate the fraction on the number line.

BUILDING UNDERSTANDING

- For students who require a challenge, have them build their skills by completing *Building Mental Strategies Skill Book Year 3* 'Locating Fractions on a Number Line', pp. 18–19. Once finished, have students create a game to help other students learn about fractions on number lines.

REFLECTION

Select from the following to suit your class and their learning outcomes:

- Ask, 'Which is bigger, $\frac{1}{4}$ or $\frac{1}{8}$? How do you know?'
- Have students stand in a circle and count by halves, e.g. $\frac{1}{2}$, 1, $1\frac{1}{2}$, 2, $2\frac{1}{2}$, 3 etc. Repeat with other fractions such as thirds, quarters and fifths.

LESSON PLAN 3

TUNING IN

FRACTIONS OF CHOCOLATE

Pose this problem: 'Yesterday I bought a block of chocolate to share. There were 24 pieces in the block. One of my friends loves chocolate and ate $\frac{1}{3}$ of the block, another ate $\frac{1}{4}$ and another ate $\frac{1}{6}$. How many pieces did they each eat and how much was left for me?' Give only a few minutes working time. Briefly discuss responses and strategies. Do not give the answer; the question will be revisited in Reflection time.

WHOLE-CLASS INTRODUCTION

FRACTIONS OF COLLECTIONS

You will need: A3 copy of BLM 72 'Cars in Our Car Park', coloured pencils

As a class, visit the school car park. Have pairs of students list the colour of 12 cars. Once back in the classroom, display an A3 copy of BLM 72 'Cars in Our Car Park'. Select one pair to share their data with the class. As they do, colour the cars in the grid. Ask, 'Look at the number of red cars. How can we work out what fraction of the whole collection is red?' Discuss, then use the arrays at the bottom of the page to simplify the fraction. For example, if there are four red cars, find an array that fits 4 as a full line – the 4 × 3 array. Colour one line in red to show that $\frac{4}{12}$ is equivalent to $\frac{1}{3}$. Repeat with another car colour. Then ask, 'What if I said one quarter of the cars were black? How could you work out how many cars this is?' Discuss and use the 4 × 3 array to answer. Explain the process of division to work out fractions of collections: $\frac{1}{4}$ of 12 = 12 ÷ 4. Ask, 'If I know the value of $\frac{1}{4}$, how could I use this to find $\frac{3}{4}$?'

INDEPENDENT TASKS

Note: Choose from Tasks 1, 2 or 3.

You will need: BLM 72 'Cars in our Car Park', *Paint,* Student Book p. 110 'Fractions of Collections'

TASK 1: CAR PARK FRACTIONS

Give students BLM 72 'Cars in Our Car Park'. Have them use their car park data from the Whole-Class Introduction to colour cars in the grid, then work out the fraction of the collection for each colour. They may use the arrays to help. Once complete, students could create a bar graph of their data.

TASK 2: INTERACTIVE TASK 

Have students use *Paint* to find different ways to represent the car park data they collected in the Whole-Class Introduction. First have them divide a rectangle into 12 (as on BLM 72 'Cars in Our Car Park') and colour the parts with all of the colours. Then have them look at one colour at a time, work out the fraction of the collection this colour is, and represent it using a partitioned shape of their choice.

TASK 3: STUDENT BOOK p. 110 *'Fractions of Collections'*

TEACHING GROUP

You will need: BLM 72 'Cars in Our Car Park', coloured counters

USING COUNTERS ON ARRAYS

- For students who require support, have them colour BLM 72 'Cars in Our Car Park' to match the data they collected in the Whole-Class Introduction. Select a car colour and have them count how many cars of that colour there are, select that number of coloured counters and arrange them on the arrays. Ask, 'Can you fill one whole line with your counters?' If so, discuss the fraction using the array as a guide. If not, have them write the fraction as $\frac{\square}{12}$.

CREATING QUESTIONS

- For students who require a challenge, have them combine two sets of car park data collected in the Whole-Class Introduction so they have 24 cars. Don't worry about specific car double-ups. Have them make a list of the fraction each colour makes, then write ten fraction questions based on their data, e.g. $\frac{3}{8}$ of 24 is ____? What fraction is the same as $\frac{12}{24}$?

REFLECTION

Select from the following to suit your class and their learning outcomes:

- Have students share their car park maths fractions, and ask, 'Which fractions were difficult to simplify? Were there any that you were unable to simplify?'
- Re-present the question from Tuning In, give students two more minutes to work on the problem, then discuss the answer (8 pieces + 6 pieces + 4 pieces = 18 eaten, 6 left). Ask, 'Has this lesson given you a strategy for solving the problem?'

Home Tasks

Select from the possible Home Tasks:

- Have students list all of the fractions from a recipe and place them in order on a number line.
- Have students find a collection of items around their house and create five fraction questions and answers about the collection, e.g. What fraction of fruit in the fruit bowl is apples?

Assessment

- Have students complete **Student Assessment p. 111**.
- Review with students **Assessment Task Card 3.27**.

During the three lessons:

- Collect any posters created by the students, making note of how they have interpreted or used their fraction knowledge.
- Collect copies of digital materials from all three lessons to add to digital portfolios. Students may wish to include a comment reflecting on their learning.
- Make a note of students' responses to the Home Tasks, and any difficulties they have had.

Recommendations for Future Learning

Specific to Student Assessment p. 111; if the student is experiencing some difficulty:

Q 1,2,4 Revise the fraction wall to look at how many parts of various fractions make a whole and to compare the size of unit fractions.

Q 3 Examine the fraction wall to help with partitioning wholes and labelling fractions. Begin by looking at fractions up to 1, and have the student follow the pattern to count further.

Q 5–6 Revise the 'Fractions of Collections' work from the Whole-Class Introduction in Lesson Plan 3.

If the student has not achieved the recommended skills for this unit:

1. See **Assessment Task Card 3.27** for specific recommendations.
2. Have the student use concrete materials to show how many parts make a whole, compare fraction size and equivalency.
3. Review *Nelson Maths: Australian Curriculum NSW Year 2* Unit 30.

If the student has achieved the recommended skills and these skills are firmly established, consider:

1. Having the student complete *Building Mental Strategies Skill Book Year 3*, pp. 4–5.
2. Moving forward to *Nelson Maths: Australian Curriculum NSW Year 4* Units 23 and 24.
3. Extending the student by working with larger fractions and collections.

Unit 28 Decimals

Number and Algebra
Fractions and decimals MA2-7NA represents, models and compares commonly used fractions and decimals

one, decimal point, tenth, whole

LESSON PLAN 1

TUNING IN

DECIMAL BRAINSTORM

Write 0.3 on the board. In their workbooks, have students brainstorm everything they know about 0.3. Then have them share their knowledge and experiences with decimals. This will enable you to gauge prior knowledge. Create a class brainstorm map on the board.

WHOLE-CLASS INTRODUCTION

INTRODUCING... TENTHS

You will need: BLM 73 'Tenths' with tenths strips cut out, magnets

Before the lesson, colour three strips from an enlarged copy of BLM 73 'Tenths' to show 0.4, 0.9 and 0.2. Use a magnet to attach the 0.4 strip to the board and ask, 'What is this a model of?' Students should recognise it as four tenths or $\frac{4}{10}$. Ask, 'How could we write this as a decimal?' Show students that the decimal value for this strip is written as 0.4. Emphasise the link between fractions and decimals. Repeat for 0.9 and 0.2. Ask, 'How could we draw a diagram of 0.3?' Discuss and have one student colour an empty strip to show 0.3. Repeat for other decimals as required.

INDEPENDENT TASKS

Note: Choose from Tasks 1, 2 or 3.

You will need: BLM 74 'Fraction–Decimal Cards', LO: *L1076 'Decimaster: match-up 1'*, Student Book p. 112 'Fraction–Decimal Snap'

TASK 1: DECIMAL SNAP

Give students BLM 74 'Fraction–Decimal Cards' (copied onto cover paper, if possible) and have them fill in the blank spaces and cut out the cards. Students shuffle the cards and place them face down in a 5 × 6 grid on the table or floor. Have them play 'Three-Way Memory' by turning over three cards at a time until they find the matching fraction, decimal and diagram.

TASK 2: INTERACTIVE TASK 

Have students work independently on computers, using LO: *L1076 'Decimaster: match-up 1'* to represent a tenth decimal as a fraction and an area model.

TASK 3: STUDENT BOOK p. 112 ***'Fraction–Decimal Snap'***

TEACHING GROUP

You will need: playdough, plastic knives, BLM 74 'Fraction–Decimal Cards'

MODELLING TENTHS

- For students who require support, have them make a flat rectangle from playdough (like a block of chocolate). Ask, 'How could you divide this into ten equal pieces?' Have students find ways to divide the shape. Once divided, take one part from one of the models, and ask, 'How much of this model do I have?' (one piece, one out of ten or one tenth) Write this on the board. Say, 'As a decimal, this is written as 0.1.' Write 0.1 on the board. Ask, 'What would 0.2 look like?' Have students model this and other decimals with their playdough.

FRACTION–DECIMAL GAMES

- For students who require a challenge, give them BLM 74 'Fraction–Decimal Cards' and have them fill in the blanks. Then have them use the cards to develop a game for one or two players. They may wish to create more cards (greater than 1) to add to their game. Have them test their games for effectiveness.

REFLECTION

Select from the following to suit your class and their learning outcomes:

- As a class, play LO: *L1076 'Decimaster: match-up 1'* on the IWB.
- Have students add any new knowledge about the decimal 0.3 to their brainstorm from Tuning In, e.g. draw a diagram, write the equivalent fraction.

LESSON PLAN 2

TUNING IN

DECIMAL BINGO!

You will need: BLM 75 'Blank Bingo Boards'

Give each student a 'Bingo' board from BLM 75 'Blank Bingo Boards'. Have students fill in the grid with decimals between zero and 1 (0.1, 0.2 up to 0.9). Call out fractions, e.g. 'Three tenths'. If a student has the matching decimal, e.g. 0.3, on their board, they cross it out. The centre 'free space' square is considered already filled. Once a student has crossed out one complete line in their grid (either vertically or horizontally), they call out 'Bingo!'

WHOLE-CLASS INTRODUCTION

DECIMALS ON THE PLACE-VALUE MAT

You will need: NTO 3.25 'Place-Value Mat with Tenths', BLM 76 'Decimal Cards'

Draw the following diagrams on the board:

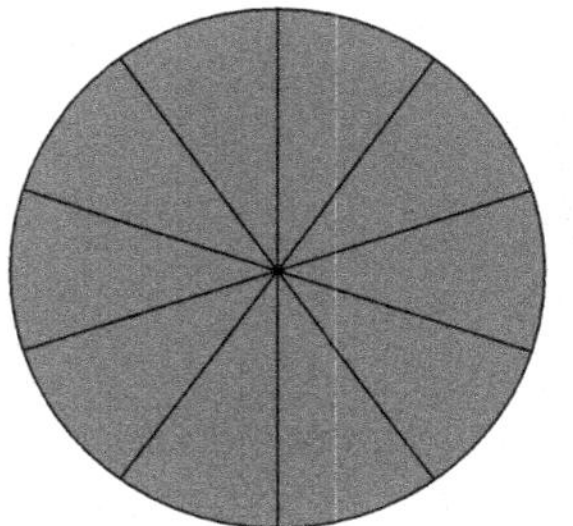
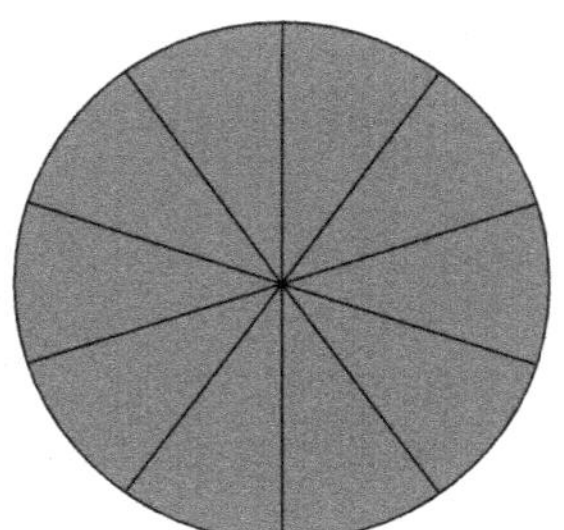
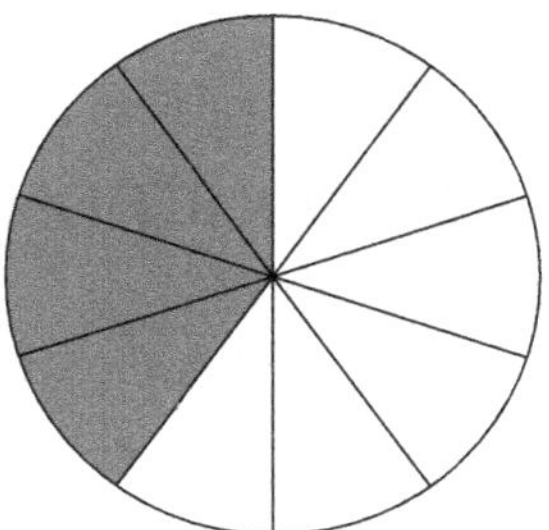

Ask, 'What number is shown here?' (2.4) Discuss. Ask, 'How would I write this on the place-value mat?' Have a student volunteer to write it. Point to the decimal point and ask, 'What is this?' Discuss, then ask, 'What does it mean if a number is after the decimal point?' Emphasise that numbers after the decimal point are the parts of a number that are smaller than 1. Ask, 'Is this number bigger or smaller than 2?' Refer students back to the diagram if required. Ask, 'Is it bigger or smaller than 3?' Discuss. Emphasise that the decimal tenths show us the parts between two numbers. Display NTO 3.25 'Place-Value Mat with Tenths' and point out the new MAB 'tenth'. Discuss its shape and size. Model 2.4 with MAB. Select a card from BLM 76 'Decimal Cards' and discuss the number (e.g. bigger than ____ but smaller than ____; has ____ tenths), write it on the place-value mat and model it with MAB. Repeat with other decimal cards as needed.

INDEPENDENT TASKS

Note: Choose from Tasks 1, 2 or 3.

You will need: MAB, playdough, plastic knives, BLM 77 'Place-Value Mat with Tenths', BLM 76 'Decimal Cards', NTO 3.25 'Place-Value Mat with Tenths', Student Book p. 113 'Zero's the Go! Decimals'

TASK 1: MAKING A TENTH MAB

Give students MAB tens and ones, playdough, a plastic knife and BLM 77 'Place-Value Mat with Tenths'. Have students write 28.4 on their place-value chart, then model whole numbers using MAB and create a way to show 'tenths' using the playdough. Have students make their tenth model accurate in relation to the MAB, i.e. $\frac{1}{10}$ of the size of a one. Once complete, have them write and model five numbers from BLM 74 'Decimal Cards'.

TASK 2: INTERACTIVE TASK 

Give students a selection of cards from BLM 76 'Decimal Cards' and have them work independently using NTO 3.25 'Place-Value Mat with Tenths' to write and model their decimal numbers.

TASK 3: STUDENT BOOK p. 113 *'Zero's the Go! Decimals'*

TEACHING GROUP

You will need: NTO 3.25 'Place-Value Mat with Tenths', MAB, playdough, plastic knives

USING THE PLACE-VALUE MAT

- For students who require support, use NTO 3.25 'Place-Value Mat with Tenths'. Start by looking at 2-digit numbers without decimal places to reinforce place value and have students model them with MAB.

Once they are confident with using the place-value mat, add a value into the tenths column. Show them an MAB one and say, 'One tenth is ten times smaller than this. Imagine this cut into ten slices.' Point out the digital tenths MAB on the IWB. Model numbers including tenths with MAB on the IWB.

MODELLING HUNDREDTHS AND THOUSANDTHS

- For students who require a challenge, discuss the values that come after tenths. Look at the whole number place-value chart to remind students that our number system works on a base 10. When moving to the left, we are multiplying each place by 10. Ask, 'What do you think might happen if we move in the other direction?' Have students discuss the idea of dividing each column by 10. Discuss the idea of tenths being ten times smaller than 1. Ask, 'What is ten times smaller than a tenth, and ten times smaller than that?' Discuss. Then have students attempt to model the values for tenths, hundredths and thousandths using playdough. This should be based on MAB.

REFLECTION

Select from the following to suit your class and their learning outcomes:

- Invite students who made models of tenths (or hundredths and thousandths) to share them with the class.
- Ask, 'What have you learned about the value of a tenth?'

LESSON PLAN 3

TUNING IN

HUMAN REACTION TIME

You will need: BLM 78 'Human Reaction Time', stopwatches or NTO 3.12 'Stopwatch'

Have students complete the instructions on BLM 78 'Human Reaction Time'.

WHOLE-CLASS INTRODUCTION

ORDERING AND NUMBER LINES

You will need: sticky notes, NTO 3.6 'Number Line'

Select ten students to write their fastest time from the Tuning In activity on a sticky note (in seconds, to one decimal place). Have them stand at the front of the class, and order themselves from fastest to slowest. Ask, 'Are you in the correct order? How do you know?' Discuss. Have students stick their sticky notes in a line on the board. Point out the first and last number, say 'All of these numbers are between ____ and ____' (using whole numbers, e.g. 5 and 12). Display NTO 3.6 'Number Line'. Write the smallest number on one end, and the largest number on the other. Have students assist in adding markers and numbers for the other whole numbers on the line. Ask, 'How many tenths are there between each of these numbers?' Students should know there are ten. Discuss the position of various tenth parts between whole numbers, e.g. 5.6, 5.9. Select a sticky note from the board, read it to the class and ask, 'Where would this be on our number line?' Select a student to draw a marker and write its value above. Repeat for other numbers on sticky notes.

INDEPENDENT TASKS

Note: Choose from Tasks 1, 2 or 3.

You will need: BLM 76 'Decimal Cards', LO: *L1078 'Decimaster plus: match-up 1,* Student Book p. 114 'Number Lines With Decimals'

TASK 1: HIGHER OR LOWER

Give pairs of students a selection of cards from BLM 76 'Decimal Cards'. Have students place the cards face down on the table. Player A selects the first card, not showing Player B. Player B is to guess the value on the card, starting by guessing whole numbers, until they have a range of 1, e.g. between 6 and 7. They then guess the decimal value. With every guess, Player A is only allowed to say 'higher' or 'lower'. Once guessed, Player B selects a card for Player A to guess. When the game is finished, students can arrange the numbers on a number line.

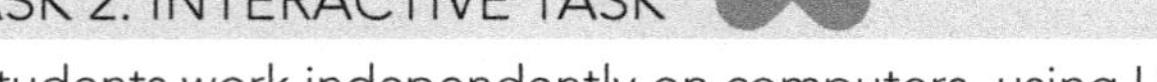

Have students work independently on computers, using LO: *L1078 'Decimaster plus: match-up 1* to represent a decimal number as a fraction and on a number line.

TASK 3: STUDENT BOOK p. 114 ***'Number Lines With Decimals'***

TEACHING GROUP

You will need: BLM 76 'Decimal Cards', chalk, BLM 78 'Human Reaction Time'

ORDER FIRST, SCALE LATER

- For students who require support, select eight decimal cards from BLM 76 'Decimal Cards' containing one whole number and one decimal, e.g. 3.1. Set these out in front of students, and read them aloud as a group. Draw a line with chalk on the floor. Have students place the smallest and biggest numbers at each end of the line. Ask, 'Which is the next smallest?' Position this just up from the smallest. Continue until all numbers are in order. Then, add scale by dividing the line and marking the whole number positions in chalk, and arranging the decimal cards. Emphasise that 0.5 equals 'half' to assist with placement.

TWO DECIMAL PLACES

- For students who require a challenge, have them work with a partner to repeat the activity on BLM 78 'Human Reaction Time' from Tuning In, with each student recording their times from five trials to two decimal places. Have them combine their results, order the numbers and place them to scale on a number line.

REFLECTION

Select from the following to suit your class and their learning outcomes:

- Give each student a decimal card and time the class to see how quickly they can order themselves from smallest to biggest without talking.
- As a class, play LO: *L1076 'Decimaster: match-up 1'* on the IWB.

Home Tasks

Select from the possible Home Tasks:

- Have students make a list of decimals found around the home, e.g. on scales, on the microwave.
- Have students develop a simple activity, like the one on BLM 76 'Human Reaction Time', that will improve with practice, e.g. a small obstacle course. Have them ask a family member to time them while they complete the activity five times and record their time to one decimal place (or two if able). Students order their numbers and place them to scale on a number line.

Assessment

- Have students complete **Student Assessment p. 115**.
- Review with students **Assessment Task Card 3.28**.

During the three lessons:

- Take photographs of students' models of tenths from Lesson Plans 1 and 2 for their digital portfolios.
- Have students write reflections on the various decimals games and whether they helped their understanding of decimals (and how) to include in their portfolios.
- Have students present their human activity homework to the class, explaining what they did and how they created the number line. Collect their homework to show the link between home and school learning.

Recommendations for Future Learning

Specific to Student Assessment p. 115; if the student is experiencing some difficulty:

Q 1 Revise the value of a tenth by modelling on a tenth strip. Make links between fractions and decimals.

Q 2 Revisit the Whole-Class Introduction activity in Lesson Plan 2. Hide the decimals column and have the student write whole numbers on the mat and once understood, add decimals.

Q 3 Have the student order whole numbers, then model the tenths on tenth strips to show which is larger.

Q 4 Revise the Whole-Class Introduction activity in Lesson Plan 3. Ensure the student understands that the decimal 0.5 is equivalent to half, and uses this knowledge to assist with placement on the number line.

If the student has not achieved the recommended skills for this unit:

1. See **Assessment Task Card 3.28** for specific recommendations.
2. Have the student work with whole numbers in any of the listed activities before moving on to decimal numbers. Scaffold with modelling and the place-value mat to reinforce values and positioning of decimals.
3. Review *Nelson Maths: Australian Curriculum NSW Year 2* Unit 30.

If the student has achieved the recommended skills and these skills are firmly established, consider:

1. Having the student complete *Building Mental Strategies Skill Book Year 3,* pp. 22–23 and pp. 24–25.
2. Moving forward to *Nelson Maths: Australian Curriculum NSW Year 4* Unit 18.
3. Extending the student in any of the listed activities by using decimal numbers to two decimal places.

Unit 29 Quadrilaterals and Symmetry

Measurement and Geometry
Two-dimensional space MA2-15MG manipulates, identifies and sketches two-dimensional shapes, including special quadrilaterals, and describes their features

angle, side, quadrilateral, parallelogram, kite, half, hexagon, irregular shapes, line of symmetry, middle, pentagon, rectangle, regular shapes, rhombus, shape, square, symmetry, trapezium, triangle

LESSON PLAN 1

TUNING IN

SHAPES

You will need: large piece of card

Have students brainstorm the 2D shapes that they know. Discuss the features of each shape. Make a shape poster on a piece of card by drawing the shapes, writing their names and how many edges and corners they have.

WHOLE-CLASS INTRODUCTION

SHAPE GUESS

You will need: bag, pattern or attribute blocks

Select one pattern or attribute block and place it in the bag without students seeing. Ask a student to put their hand in the bag and try to work out which shape is hidden. Encourage the student to try to identify the shape's features by feeling the edges and corners. Have the student describe what they can feel to the other students. When the student has named a shape, reveal the shape in the bag. Repeat with a different shape.

INDEPENDENT TASKS

Note: Choose from Tasks 1, 2 or 3.

You will need: BLM 79 '2D Shapes', attribute or pattern blocks (optional), NTO 3.26 '2D Shapes', Student Book p. 116 'Special Quadrilaterals'

TASK 1: COMPARE AND CONTRAST

Have students select two cards from BLM 79 '2D Shapes' and record the features of the two shapes, i.e. how many sides and angles, length of sides. Have students then compare and contrast the two 2D shapes by recording how they are similar and different. If students find this difficult using the pictures, give them attribute blocks or pattern blocks to use.

TASK 2: INTERACTIVE TASK 

Have students work with NTO 3.26 '2D Shapes', select a 2D shape and then select 'notes'. In the 'notes' box, students type a description of the shape. They continue with other shapes, then ask a partner to check if their descriptions are accurate. Consider focusing on special quadrilaterals and discuss the similarities and differences between different types of quadrilaterals.

TASK 3: STUDENT BOOK p. 116 *'Special Quadrilaterals'*

TEACHING GROUP

You will need: BLM 79 '2D Shapes', craft sticks

LOOKING FOR SHAPES

- For students who require support, have them select cards from BLM 79 '2D Shapes' and search for things around the room that are the same shapes as the shapes on the cards. Students draw pictures of the items they find and then describe them according to their features. Ask, 'Which are the most common shapes?'

SHAPE CHALLENGE

- For students who require a challenge, give them three craft sticks each. Ask, 'How many different shapes can you make with three craft sticks?' Have students record the shapes they are able to make, then give them an extra craft stick and ask, 'How many different shapes can you make with four craft sticks?' Talk about the shapes that students record and discuss how some are regular (such as a square) and others are irregular. Repeat with five and six craft sticks.

REFLECTION

Select from the following to suit your class and their learning outcomes:

- Have students explain the features of different 2D shapes. Ask, 'How do the sides and angles help us to describe shapes?'
- Play a shape description game with students by describing the features of a 2D shape, e.g. 'It has four edges and four corners.' Have students guess the shape you are describing.

LESSON PLAN 2

TUNING IN

BODY SHAPE

You will need: pieces of butcher's paper taped together

Select a student to lie on top of the sheets of butcher's paper and trace around their body. Then draw a line down the middle of the outline. Have students discuss what they notice. Talk about how the shape of the body is symmetrical, which means that one side is a mirror image of the other. Have students work with a partner, where a student traces around one half of their partner's body (lengthwise). They then try to draw the other half so that it is symmetrical. Have students swap roles.

WHOLE-CLASS INTRODUCTION

IS IT SYMMETRICAL?

You will need: BLM 80 'Is it Symmetrical?', mirror

Enlarge and cut out the shapes on BLM 80 'Is it Symmetrical?' Present the square to students and ask, 'Is this shape symmetrical? How could we find out?' Discuss students' ideas. Fold the square in half to show how the two sides are the same. Ask, 'Are there any other ways we could fold the shape for it to be symmetrical?' Discuss how the line in the middle of a symmetrical shape is called the line of symmetry. Repeat with other shapes and identify the ones that are or are not symmetrical. Alternatively, demonstrate how placing a mirror on the line of symmetry will show if the shape is symmetrical.

INDEPENDENT TASKS

Note: Choose from Tasks 1, 2 or 3.

You will need: digital camera, *Word*, mirrors, Student Book p. 117 'Symmetry'

TASK 1: SYMMETRICAL OBJECTS

Have students search for symmetrical objects around the room, e.g. table, book, chair. Students make a list of the symmetrical objects they find, draw a picture of them and then draw in the lines of symmetry. Alternatively, they could take photos of the symmetrical objects using a digital camera, print the photos and draw in the lines of symmetry.

TASK 2: INTERACTIVE TASK 

Have students work on computers, using *Word* to make a poster of symmetrical shapes. Have them use the shape tool to find and select symmetrical shapes and then arrange them on a page, under the heading 'Symmetrical Shapes'.

TASK 3: STUDENT BOOK p. 117 *'Symmetry'*

TEACHING GROUP

You will need: BLM 80 'Is it Symmetrical?', BLM 81 'Alphabet Cards'

WHICH SHAPES ARE SYMMETRICAL?

- For students who require support, give them BLM 80 'Is it Symmetrical?' and have them identify the symmetrical shapes. Students find the lines of symmetry by placing a mirror on the middle of the shapes or by cutting out and folding the shapes.

LETTERS AND WORDS

- For students who require a challenge, give them cards made from BLM 81 'Alphabet Cards'. Students look at each letter and work out if it is symmetrical, either vertically or horizontally. Have them sort the symmetrical letters into two categories: vertically symmetrical and horizontally symmetrical. Encourage them to fold the letter cards if necessary. To further extend students, have them identify and record symmetrical words, e.g. COD (horizontally symmetrical).

REFLECTION

Select from the following to suit your class and their learning outcomes:

- Have students reflect on the Independent Tasks and Teaching Group activities. Have them explain what it means if something is symmetrical. Encourage them to use objects or drawings to help their explanation.
- Walk around the school and have students identify examples of symmetry. Have them look at buildings and items from nature, e.g. leaves and flowers. Ask, 'How do you know they are symmetrical?'

LESSON PLAN 3

TUNING IN

SYMMETRY IN ART

You will need: examples of Aboriginal rock carvings or art

Show students examples of Aboriginal art or rock carvings. Have them identify examples of symmetry in the artwork. Ask, 'Why do you think there are so many examples of symmetry in the artwork'?

WHOLE-CLASS INTRODUCTION

MULTIPLE LINES OF SYMMETRY

You will need: large cardboard shapes, e.g. rectangle, triangle, arrow

Show students the large cardboard rectangle. Ask, 'Is this shape symmetrical?' Have students explain how they know the shape is symmetrical. Fold the rectangle in half to show the line of symmetry. Then fold the rectangle in half the other way to identify the other line of symmetry. Discuss how some shapes have more than one line of symmetry. Repeat with the other cardboard shapes. Ask, 'How many lines of symmetry do these shapes have?'

INDEPENDENT TASKS

Note: Choose from Tasks 1, 2 or 3.

You will need: BLM 79 '2D Shapes', mirror, BLM 10 'Grid Paper', NTO 3.26 '2D Shapes', Student Book p. 118 'Lines of Symmetry'

TASK 1: HOW MANY LINES OF SYMMETRY?

Give students cards made from BLM 79 '2D Shapes' and have them determine the number of lines of symmetry in each shape. If necessary, they can use a mirror to find the lines of symmetry. Students sort the shapes into groups based on how many lines of symmetry, e.g. no lines, one line, two lines. Then give students BLM 10 'Grid Paper' and have them draw and cut out other shapes for each group.

TASK 2: INTERACTIVE TASK 

Have students work with a partner, using NTO 3.26 '2D Shapes' to select shapes to design a symmetrical picture. To challenge students further, have them design a picture that has more than one line of symmetry. Students ask a partner to identify the lines of symmetry in their picture.

TASK 3: STUDENT BOOK p. 118 *'Lines of Symmetry'*

TEACHING GROUP

You will need: square paper, ruler, pattern blocks

SYMMETRICAL SNOWFLAKES

- For students who require support, have them make a snowflake using a piece of square paper. Have them fold their paper into quarters and then cut out shapes and patterns, ensuring they cut through all four layers of paper. Students open their snowflake and discuss what they notice. Ask, 'How many lines of symmetry does your snowflake have?' Have them draw the lines of symmetry on their snowflake using a ruler. Discuss the strategies students used to work out the lines of symmetry.

SYMMETRY WITH PATTERN BLOCKS

- For students who require a challenge, present them with the following pattern made from pattern blocks: square, rectangle, square, rectangle, square. Ask, 'Where are the lines of symmetry in this pattern?' Discuss how this pattern has two lines of symmetry. Have students use pattern blocks to make their own patterns or designs and then identify how many lines of symmetry they have.

REFLECTION

Select from the following to suit your class and their learning outcomes:

- Have students explain the strategies they can use to identify the number of lines of symmetry in a shape, pattern or design.
- Have students reflect on Independent Tasks, Task 2. Discuss the strategies they used to design a symmetrical picture.

Home Tasks

Select from the possible Home Tasks:

- Have students identify 2D shapes around the home and describe their features.
- Have students find symmetrical objects and shapes around the home, and identify the line of symmetry.

Assessment

- Have students complete **Student Assessment p. 119**.
- Review with students **Assessment Task Card 3.29**.

During the three lessons:

- Observe which students are able to identify symmetrical shapes and objects.
- Make note of students who completed the scaffolding tasks or the more challenging activities of the Teaching Groups.
- Review Student Book pages and make notes of areas of difficulty.

Recommendations for Future Learning

Specific to Student Assessment p. 119; if the student is experiencing some difficulty:

Q 1 Revise 2D shapes and their features. Provide opportunities for the student to describe 2D shapes by identifying the number of edges and corners.

Q 2–3 Review what 'symmetrical' means and have the student explore shapes that can be folded in half to produce two identical halves.

Q 4 Revise how some shapes have more than one line of symmetry. Encourage the student to use mirrors or to fold cut-outs of shapes to identify the lines of symmetry.

If the student has not achieved the recommended skills for this unit:

1. See **Assessment Task Card 3.29** for specific recommendations.
2. Have the student fold cut-outs of shapes or place mirrors along the line of symmetry to recognise how shapes are symmetrical.
3. Review *Nelson Maths: Australian Curriculum NSW Year 2* Unit 8.

If the student has achieved the recommended skills and these skills are firmly established, consider:

1. Moving forward to *Nelson Maths: Australian Curriculum NSW Year 4* Unit 8.
2. Providing opportunities for the student to explore tessellation with 2D shapes.

Unit 30 Volume and Capacity

Measurement and Geometry
Volume and capacity MA2-11MG measures, records, compares and estimates volumes and capacities using litres, millilitres and cubic centimetres

capacity, cups, fill up, full, holds, litre, millilitre, volume, cubic centimetre (cm^3)

LESSON PLAN 1

TUNING IN

HOW MANY UNIFIX?

You will need: large container full of Unifix blocks (or any other uniform units)

Present the filled container to the class, and ask, 'How many Unifix blocks are in this container?' Allow students to make their predictions. The 'winner' is the student/s closest to the actual amount.

WHOLE-CLASS INTRODUCTION

WHICH CONTAINER HOLDS THE MOST?

You will need: three boxes similar in size but with different dimensions, uniform units (e.g. marbles, MAB ones, Unifix blocks)

Present the three boxes to the class, and ask, 'Which box do you think will hold the most?' Invite students to share their predictions and explain their reasoning. Then ask, 'How could we test to see which holds the most? What could we use?' Lead discussion that a uniform unit is needed. Select students to fill the boxes with uniform units and count how many were used. Compare and discuss results.

INDEPENDENT TASKS

Note: Choose from Tasks 1, 2 or 3.

You will need: marbles (or other uniform unit), LO: *L1994 'Squirt: two containers: level 1'*, Student Book p. 120 'Tin Cans and Egg Cups'

TASK 1: THE BIGGEST SHOE

Have students estimate how many marbles it will take to fill their shoe. Students work in small groups to find out how much each student's shoe holds, then order the shoes in their group from smallest to biggest. Have students look for something else in the room that holds the same amount of marbles.

TASK 2: INTERACTIVE TASK

Have students work independently on computers, using LO: *L1994 'Squirt: two containers: level 1'* to examine the relationships between capacities of containers of the same shape but different size. They fill containers and squirt liquids between them to establish the proportional relationship.

TASK 3: STUDENT BOOK p. 120 *'Tin Cans and Egg Cups'*

TEACHING GROUP

You will need: tennis balls, different sized containers, small whiteboard, large container, school sink, stopwatch, containers for measuring (e.g. cups, cans)

TENNIS ANYONE?

- For students who require support, give them tennis balls and different sized containers. As a group, look at one container and ask, 'How many tennis balls will fit in this container?' Have students make predictions and discuss. Test how many balls will fit; put the small whiteboard on top of the container to ensure balls are not higher than the container. Discuss outcomes. Have pairs of students repeat the process with other containers.

WATER FROM A TAP

- For students who require a challenge, have them work as a group to place a large container in the sink and turn the tap on full for exactly ten seconds. Have them find a way to measure the amount of water released by the tap in ten seconds. They may use any measure they like, e.g. cups, cans. Then, have them repeat the activity for 20 seconds – was it exactly double the amount? And one minute – was it six times the amount? Have them collate and explain results.

REFLECTION

Select from the following to suit your class and their learning outcomes:

- As a class, play LO: *L1994 'Squirt: two containers: level 1'* on the IWB.
- Ask, 'When was it difficult to estimate the capacity of an object? What strategies did you use?'

LESSON PLAN 2

TUNING IN

ORDERING CAPACITY

You will need: five containers of varying shape and size filled with water and labelled A to E

Present the five containers and ask students to order them from smallest to largest capacity. Remind students that capacity is how much a container holds. Have students record their predictions, and discuss and collate to form a class order. Ask, 'How could we measure the amount of water in each to see if our predictions are accurate?' Allow students to provide responses, but do not measure the water.

WHOLE-CLASS INTRODUCTION

MEASURING JUGS

You will need: measuring jugs, water-filled containers from Tuning In

Present measuring jugs and ask, 'What do we use these for?' Discuss the use of measuring jugs inside and outside the home. Look at the scales on the jugs. Invite students to explain what they are and how they work. Use the jugs to measure the amount of water in each of the containers from Tuning In, one at a time, discussing with each measure whether the predictions were accurate. Invite students to read the scale on the jug for each measurement.

INDEPENDENT TASKS

Note: Choose from Tasks 1, 2 or 3.

You will need: ten 1 L transparent containers filled with different amounts of water and labelled A to J, BLM 83 'Investigating Millilitres', measuring jugs, *PowerPoint,* Student Book p. 121 'Measuring Jugs'

TASK 1: HOW MANY MILLILITRES?

Place the ten containers around the classroom. Have students work in pairs, moving around the room and estimating the amount of water in each container. They record their estimates on BLM 83 'Investigating Millilitres'. Once students have completed their estimations, each pair uses a measuring jug to measure the amount of water in one of the containers. Class results are combined and recorded on the BLM. Have students compare their estimations and actual amounts.

TASK 2: INTERACTIVE TASK 

Have students create a *PowerPoint* presentation of everyday use of capacity. Their presentation should cover millilitres and litres, and have pictures of items measured in each category.

TASK 3: STUDENT BOOK p. 121 *'Measuring Jugs'*

TEACHING GROUP

You will need: five clear plastic cups filled with different amounts of water and labelled A to E, measuring jugs, containers, sand/rice, kitchen scales

ORDERING AND MEASURING

- For students who require support, present them with the five cups and have them write them in order from least to most full. Then, have them measure the amount of water in each by taking turns to tip the water carefully into a measuring jug and reading the scale to find the amount in millilitres. Record and discuss results.

MASS AND CAPACITY

- For students who require a challenge, give them containers and sand/rice. Have them predict whether the containers will have a greater mass or capacity. Have them measure both mass (with scales) and capacity (with measuring jug), then compare and discuss results.

REFLECTION

Select from the following to suit your class and their learning outcomes:

- Have students share any strategies for predicting the amount of water and their experiences using the jugs to measure.
- Have students who did the 'Mass and Capacity' activity in the Teaching Group share their knowledge about the relationship between mass and capacity with the class.

LESSON PLAN **3**

TUNING IN

WHAT DO WE MEAN BY VOLUME?

You will need: two balls of different sizes

Present students with the question, 'Where have you heard the word volume?' The most common understanding of the term is likely to be in connection with levels of loudness. Show the two balls and explain that the volume of each is different. Ask, 'What do we mean by volume this time?' Identify that the larger ball has a greater volume because it takes up more space than the smaller one.

WHOLE-CLASS INTRODUCTION

WE ALL HAVE VOLUME

Remind students of what is meant by volume in connection with volume and capacity. If appropriate, explain that the difference between the two is that capacity is the amount that can be poured into a container whereas volume is the amount of space something takes up. Elaborate by explaining that we all take up space so we all have volume. Ask students to name something in the school that has a larger volume than themselves and something that has a smaller volume.

INDEPENDENT TASKS

Note: Choose from Tasks 1, 2 or 3.

You will need: interlocking centimetre cubes, NTO 3.28 'Multilink Blocks', Student Book p. 122 'Volume'

TASK 1: MAKING INTERLOCKING CUBE MODELS

Review the concept of volume as the amount of space that something takes up. Give students (individually or in pairs) four interlocking centimetre cubes and ask them to join them together. Have students share the way they have made their models. Establish the volume as 4 cm^3. If each model is a line of four cubes, ask if there is another way the cubes can be joined. Students share the models and say what the volume is. Repeat with different numbers of cubes.

TASK 2: INTERACTIVE TASK 

Use NTO 3.28 'Multilink Blocks' to build models pictorially to show that models with the same volume do not necessarily have the same dimensions.

TASK 3: STUDENT BOOK p. 122 *'Volume'*

TEACHING GROUP

You will need: interlocking centimetre cubes, NTO 3.28 'Multilink Blocks', Student Book p. 122 'Volume'

MODELLING CUBES

- For students who require support, make available a box of interlocking centimetre cubes. Establish that each cube has a volume of one cubic centimetre (1 cm^3). Use NTO 2.28 'Multilink Blocks' and jointly construct a model using six blocks. Ask students to copy the model using interlocking cubes. Have students identify the volume of the model. Repeat with different numbers of blocks and cubes.

RECTANGULAR PRISMS

- For students who require a challenge, build a model using interlocking centimetre cubes that is 4 cubes by 2 cubes. Ask students to predict the volume if a second layer is added (or further layers). Repeat with rectangular models of different dimensions.

REFLECTION

Select from the following to suit your class and their learning outcomes:

- Say, 'I have 10 cubes that I am going to join together. What will be the volume of the finished model?'
- Say, 'Sam and Sara each used cubes to build a model. Both models were different but each had a volume of 12 cm^3. What might the models have looked like?'

Home Tasks

Select from the possible Home Tasks:

- Have students make a list of containers in their home that have a capacity of less than 1 L, equal to 1 L and more than 1 L.
- Have students investigate the shapes of boxes and containers. What is the most common shape?

Assessment

- Have students complete **Student Assessment p. 123**.
- Review with students **Assessment Task Card 3.30**.

During the three lessons:

- Take photos of students working with the various informal and formal measuring devices to add to their digital portfolios. Students could be invited to write a reflection on the activity and their learning.
- Have students write down five things they have learned about volume and capacity in this unit and collect these to show understanding.
- Have students make as many different interlocking-cube models as they can that have a volume of 12 cm^3. Make a labelled display of the results.

Recommendations for Future Learning

Specific to Student Assessment p. 123; if the student is experiencing some difficulty:

Q 1–2 Revise content from Lesson Plan 2, where litres and millilitres are introduced and discussed. Provide examples of containers that would fit litres and millilitres, which clearly show litres to be a larger measure. Have the student think about shopping catalogues or items in their kitchen.

Q 3 Present containers/bottles that have sizes similar to those listed and allow the student to compare their sizes. Emphasise the relationship between litres and millilitres.

Q 4 Remind the student of the activities in Lesson Plan 1 – this is the same concept, but using a formal measure. If needed, provide the student with concrete materials and allow them to fill larger containers with water using a 250 mL measure.

Q 5 Review the difference between capacity and volume and look for opportunities in the classroom environment to reinforce the concepts.

If the student has not achieved the recommended skills for this unit:

1. See **Assessment Task Card 3.30** for specific recommendations.
2. Have the student further their understanding of capacity by comparing two items before moving on to ordering collections of items.
3. Have the student continue to use hands-on materials, both formal and informal, to compare capacity and to recognise volume as the amount of space that something occupies.
4. Review *Nelson Maths: Australian Curriculum NSW Year 2* Unit 20.

If the student has achieved the recommended skills and these skills are firmly established, consider:

1. Having the student extend their knowledge of litres by working with larger amounts.
2. Moving forward to *Nelson Maths: Australian Curriculum NSW Year 4* Unit 5.
3. Having the student investigate shortcuts for finding the volume of a rectangular prism.

Unit 31 3D Objects

Measurement and Geometry

Three-dimensional space MA2-14MG makes, compares, sketches and names three-dimensional objects, including prisms, pyramids, cylinders, cones and spheres, and describes their features

2D shape, 3D object, circle, cone, corner, cube, cylinder, edge, face, hexagon, prism, pyramid, rectangle, rectangular prism, sphere, square, tetrahedron, triangle, triangular prism

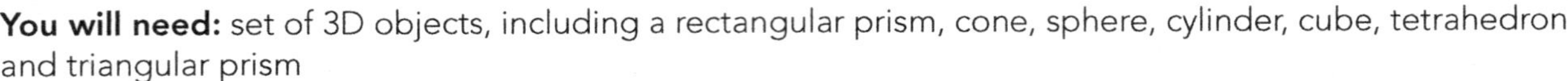

LESSON PLAN 1

TUNING IN

3D OBJECTS

You will need: set of 3D objects, including a rectangular prism, cone, sphere, cylinder, cube, tetrahedron and triangular prism

Present students with a set of 3D objects and have them identify the ones with which they are familiar. Select one object at a time and have students talk about what it looks like. Ask, 'Where have you seen this 3D object before?' Record the names of the 3D objects on a poster for students to refer to throughout the lesson.

WHOLE-CLASS INTRODUCTION

DESCRIBE THE 3D OBJECT

You will need: set of 3D objects

Present students with a cube. Identify and count the faces, corners and edges. Discuss how the features of the cube help us to describe it. Ask, 'How would you describe the cube to somebody? What 2D shapes can you see?' Have students identify the shape of the faces. Select other 3D objects and have students identify and count the faces, corners and edges. Then have them describe the object to a partner.

INDEPENDENT TASKS

Note: Choose from Tasks 1, 2 or 3.

You will need: BLM 84 '3D Object Templates 1', BLM 85 '3D Object Templates 2', LO: *L1068 'Face painter: finding faces 1'*, paper, Student Book p. 124 '3D Objects'

TASK 1: MAKE AND DESCRIBE

Give students BLM 84 '3D Object Templates 1' and BLM 85 '3D Object Templates 2'. Have students select a 3D object template, cut it out and fold it to make a 3D object. Students write a description of their 3D object by recording the number of faces, corners and edges, and describing the shape of the faces. Have students write descriptions for other 3D objects by either making different 3D objects or swapping shapes with a partner.

TASK 2: INTERACTIVE TASK 

Have students work on computers, using LO: *L1068 'Face painter: finding faces 1'* to identify the faces of 3D objects. Give students a sheet of paper and have them record the number of faces, corners and edges for each 3D object they explore.

TASK 3: STUDENT BOOK p. 124 ***'3D Objects'***

TEACHING GROUP

You will need: digital camera, playdough, plastic knives

3D OBJECT HUNT

- For students who require support, have them search for and photograph examples of 3D objects around the room. Students then sort the objects into the following categories: cubes, triangular prisms, cones, spheres, tetrahedrons, cylinders and rectangular prisms. Encourage students to make other categories if necessary. Have students explore the features of the objects in each category.

CROSS SECTIONS

- For students who require a challenge, have them use playdough to make models of 3D objects, then identify and record the shapes of the faces. Explain to students if they cut a 3D object in half they will be able to see the cross-section. Have students predict what shape the cross-sections will be, then cut the shapes in half and discuss what they notice. Ask, 'Were the cross-sections the same shape as any of the faces of the 3D object?'

REFLECTION

Select from the following to suit your class and their learning outcomes:

- Have students explain the features of different 3D objects. Ask, 'How do the faces, edges and corners help us to describe 3D objects?'
- Have students reflect on the 3D object hunt in the Teaching Group. Discuss which 3D objects were more common around the classroom. Ask, 'Why do you think some 3D objects are used more often than others?'

LESSON PLAN 2

TUNING IN

LOOK AT THE PYRAMIDS

You will need: pictures of pyramids (e.g. from the internet or *ClipArt*)

Show students pictures of pyramids. Have them discuss the images and talk about where pyramids are found and how they may have been built. Ask, 'What 2D shapes can you see when you look at a pyramid?' Have students describe the pyramids.

WHOLE-CLASS INTRODUCTION

DESCRIBE THE PYRAMIDS

You will need: pyramids from a set of 3D objects

Present students with pyramids from a set of 3D objects. Have students look at the pyramids and discuss what they notice. Ask, 'How are the pyramids the same? How are the pyramids different?' Establish that pyramids can have different shaped bases, but all the other faces are triangles. Discuss how pyramids are named according to their base, e.g. a hexagonal pyramid has a hexagonal base. Make a list of different types of pyramids: square pyramid, rectangular pyramid, hexagonal pyramid, triangular pyramid.

INDEPENDENT TASKS

Note: Choose from Tasks 1, 2 or 3.

You will need: newspapers, masking tape, *PowerPoint*, pyramids from a set of 3D objects, Geoshapes or Polydron, digital camera, Student Book p. 125 'Pyramids'

TASK 1: PYRAMID FEATURES

Have students refer to the list of pyramids from the Whole-Class Introduction. Have them select a pyramid from the list and construct it by rolling up pieces of newspaper to form the edges of the pyramid and using masking tape to stick the edges together, i.e. it will form a skeleton shape. Have students make a list of the features of their pyramids and record the number of faces, edges and corners of each type of pyramid.

TASK 2: INTERACTIVE TASK 

Have students make a *PowerPoint* presentation about pyramids, describing the features of pyramids and including pictures. Encourage them to look at pyramids from a set of 3D objects if necessary. Students can make models of pyramids using Geoshapes or Polydron, take photos of them and add the photos to their presentation.

TASK 3: STUDENT BOOK p. 125 *'Pyramids'*

TEACHING GROUP

You will need: pyramids from a set of 3D objects, butcher's paper, paper bag, pipe cleaners or toothpicks and modelling clay

TRACE THE SHAPE

- For students who require support, give them pyramids from a set of 3D objects. Have them select a pyramid and place it on a sheet of butcher's paper. Students trace around the edges of their shape to identify the 2D shape. Have students trace around all faces of the pyramid. Discuss what students notice.

HIDDEN PYRAMID

- For students who require a challenge, place a pyramid from a set of 3D objects in a paper bag so that it is hidden. Describe the shape of its base, and the shape and number of its faces. Give students pipe cleaners or toothpicks and modelling clay, and have them build a model of a pyramid that matches your description. When they have finished, take the pyramid out of the bag and compare it to the students' models. Then have students work with a partner, taking turns to describe the features of a hidden pyramid while their partner constructs the shape with the materials.

REFLECTION

Select from the following to suit your class and their learning outcomes:

- Have students reflect on the Independent Tasks and Teaching Group activities. Have them share what they have learned about pyramids and describe the features of a pyramid.
- Present students with models of different types of pyramids. Ask, 'What is similar and different about these pyramids? How are pyramids different from other 3D objects?'

LESSON PLAN 3

TUNING IN

WHAT IS A PRISM?

You will need: prisms from a set of 3D objects

Present students with prisms from a set of 3D objects and have them describe their features. Ask, 'What do these 3D objects have in common?' Establish that they are all prisms because they have two faces that are same shape (one at each end of the prism) and then all the other faces are the same shape (and are usually rectangles or squares). Show students the triangular prism and have them identify the identical triangular faces at each end of the prism and then the other identical rectangular faces. Repeat with the other prisms.

WHOLE-CLASS INTRODUCTION

PRISMS AND PYRAMIDS

You will need: pyramids and prisms from a set of 3D objects

Present students with prisms and pyramids from a set of 3D objects. Have students sort the shapes into two categories: prisms and pyramids. Ask, 'How did you know which category a shape went into?' Discuss the features of the shapes in each group. Ask, 'What is similar and different about these groups of 3D objects?'

INDEPENDENT TASKS

Note: Choose from Tasks 1, 2 or 3.

You will need: playdough or modelling clay, digital camera, Student Book p. 126 'Prisms and Pyramids'

TASK 1: MAKE 3D OBJECTS

Give students playdough or modelling clay. Have them make models of different types of prisms, e.g. rectangular prisms, triangular prisms, hexagonal prisms. Have students then record the number of edges, corners and faces of each prism and also the shape of its faces.

TASK 2: INTERACTIVE TASK 

Have students go on a prism and pyramid hunt around the classroom or school and take photos of the prisms and pyramids they find. Students print their photos and then record the numbers of sides, faces and corners of each prism or pyramid.

TASK 3: STUDENT BOOK p. 126 ***'Prisms and Pyramids'***

TEACHING GROUP

You will need: BLM 86 'Prisms and Pyramids', prisms and pyramids from a set of 3D objects

GUESSING GAME

- For students who require support, have them play a prism and pyramid guessing game with a partner. Students make cards using the prisms and pyramids on BLM 84 'Prisms and Pyramids'. Have one student select one of the cards, without their partner seeing. They describe the features of the object to their partner. If their partner is able to guess which prism or pyramid it is, they receive one point. If necessary, give students prisms and pyramids from a set of 3D objects to help them identify the correct prism or pyramid. Students then swap roles.

COMPARE AND CONTRAST

- For students who require a challenge, give them BLM 84 'Prisms and Pyramids'. Have them select two objects and describe their features, i.e. the number of faces, corners and edges, and the shape of the faces. Have students then compare and contrast the two objects by recording how they are similar and how they are different. If students find this difficult using the pictures, give them a set of 3D objects to use.

REFLECTION

Select from the following to suit your class and their learning outcomes:

- Have students explain the features of a prism. Ask, 'How are all prisms similar and different?'
- Have students reflect on the object hunt in Independent Tasks, Task 2. Ask, 'Did you find more prisms or pyramids on your object hunt? Why do you think this was?'

Home Tasks

Select from the possible Home Tasks:

- Have students find examples of 3D objects around the home. Ask, 'Which is the most common 3D object in your house?'
- Present students with real-life 3D objects, e.g. book, box, can, ball. Have students describe their features and explain how they are similar or different.

Assessment

- Have students complete **Student Assessment p. 127**.
- Review with students **Assessment Task Card 3.31**.

During the three lessons:

- Observe which students are able to make models of 3D objects (including prisms and pyramids) and describe their features.
- Make note of students who completed the scaffolding tasks or the more challenging activities of the Teaching Groups.
- Review Student Book pages and make notes of areas of difficulty.

Recommendations for Future Learning

Specific to Student Assessment p. 127; if the student is experiencing some difficulty:

Q 1 Revise the features of 3D objects. Have the student identify and count the faces, edges and corners of 3D objects. Have the student practise describing 3D objects by referring to their features.

Q 2 Revise how the faces of 3D objects are 2D shapes. Present the student with models of 3D objects (including prisms and pyramids) and have them identify the shape of their faces. Encourage the student to trace around the faces to recognise the 2D shapes.

Q 3 Revise the features of prisms and pyramids. Present the student with prisms and pyramids from a set of 3D objects. Have them sort them into two groups and then describe the objects in each group.

If the student has not achieved the recommended skills for this unit:

1. See **Assessment Task Card 3.31** for specific recommendations.
2. Have the student make models of 3D objects using Geoshapes or Polydron or with cardboard and tape. Have the student identify the features of these 3D objects.
3. Review *Nelson Maths: Australian Curriculum NSW Year 2* Unit 26.

If the student has achieved the recommended skills and these skills are firmly established, consider:

1. Moving forward to *Nelson Maths: Australian Curriculum NSW Year 4* Unit 8.
2. Providing opportunities for students to explore other 3D objects, e.g. pentagonal prisms and hexagonal prisms.

Unit 32 Money

Number and Algebra

Addition and subtraction (money) MA2-5NA uses mental and written strategies for addition and subtraction involving two-, three-, four- and five-digit numbers

addition, cents, change, coins, currency, difference, dollars, notes, subtraction, total, rounding, rounded total

LESSON PLAN 1

TUNING IN

THE VALUE OF THREE

Pose the question: 'Yesterday at the shops I paid for an item with three coins. How much might the item have cost? What could it have been?' Discuss responses and note whether students have a reasonable idea of how much things cost to gauge their experience with money.

WHOLE-CLASS INTRODUCTION

OUR COINS

You will need: NTO 3.27 'Australian and International Coins', poster paper

Display NTO 3.27 'Australian and International Coins'. Ask, 'What coins do we use in Australia?' Discuss the value, size and colour of the coins, selecting each coin on the IWB as they are discussed. List this information on a poster to display in the classroom. Discuss the value of $1. Ask, 'How many cents is $1 worth?' Students should recognise that it is worth 100 cents. If not, ask, 'How many 50-cent pieces do I need to make $1?' This should help students gain understanding of the value of $1. Give students the opportunity to add coin values by displaying sets of two coins and asking, 'What is the total value?' Further understanding by displaying larger sets of coins. Select the 'Show international coins' button and select a coin from the list. Ask, 'What is this? Have you seen one of these before?' Discuss the coin and where it is from. Ask, 'Can we use these coins in Australia?' Discuss, so that students realise that each country has its own coins and they can't be used in other countries as they hold different values. Discuss the similarities and differences between the selection of international coins and Australian coins.

INDEPENDENT TASKS

Note: Choose from Tasks 1, 2 or 3.

You will need: BLM 88 'Coin Challenges', BLM 87 'Coins' (or play money if available), NTO 3.27 'Australian and International Coins', access to the internet, Student Book p. 128 'Coins, Coins, Coins'

TASK 1: COIN CHALLENGES

Give students a selection of cards made from BLM 88 'Coin Challenges' and a copy of BLM 87 'Coins' and have students complete the challenges on the cards.

TASK 2: INTERACTIVE TASK 

Have students work independently on computers, using NTO 3.27 'Australian and International Coins' to use Australian coins to make $1 in five different ways. Then have them explore coins in the international coins list. Have them select their favourite coin and research coins from this country on the internet. Compare the appearance and value of their coins to Australian coins.

TASK 3: STUDENT BOOK p. 128 *'Coins, Coins, Coins'*

TEACHING GROUP

You will need: BLM 87 'Coins' (or play money), BLM 88 'Coin Challenges'

MAKING $1

- For students who require support, give them coins cut out from BLM 87 'Coins'. Put two 50-cent pieces together and ask, 'How much is this worth?' Students should recognise this as $1; if not, emphasise that $1 is 100 cents, and 50 + 50 = 100. Then, ask students, 'How many different ways can you make $1?' Have students use the coins to make $1 as many different ways as they can.

COMPETITIVE CHALLENGES

- For students who require a challenge, have them work in pairs with three cards from BLM 88 'Coin Challenges'. Place the cards face down. One card is turned over and the pair races to see who can complete the challenge first. The best of three challenges wins. Once complete, have students create more difficult coin challenges for their partner.

REFLECTION

Select from the following to suit your class and their learning outcomes:

- Ask, 'Why do you think coins have different sizes, shapes and colours?'
- Ask, 'Why do you think our coins have the values they do? Why don't we have coins worth, e.g. 30 cents, 40 cents, 60 cents?'

LESSON PLAN 2

TUNING IN

THE PRICE IS RIGHT!

You will need: toy catalogues (with prices that require rounding)

Choose six toys from a toy catalogue with a variety of prices – one or two small/cheap items, two or three mid-range items and one or two more expensive items. Cover the prices and display the advertised toys on the board. Do not place in order of price. Introduce each item, pointing out key selling points, brand and so on, e.g. 'Brought to you by [brand name], this toy is ...' Have students order them from cheapest to most expensive. Once completed, invite a student to move the advertisements into their predicted order. Reveal the actual prices of the toys one by one. You may allow the student to change the order of unrevealed items as the prices of others are shown. Keep the information from this activity for the Whole-Class Introduction.

WHOLE-CLASS INTRODUCTION

THE PRICE IS RIGHT – ALMOST!

You will need: information from the Tuning In activity

Point out a price from the toy catalogue that requires rounding and ask something like, 'What do you notice about the price of the teddy bear? Is it possible to pay for it with an exact number of coins? Why not?' Ask what other items need to be similarly rounded. Depending on the ability level and interest of the group, students may also like to reflect on why shops sometimes sell items at prices that need to be rounded.

INDEPENDENT TASKS

Note: Choose from Tasks 1, 2 or 3.

You will need: toy catalogues, A3 paper, *Excel*, BLM 89 'Supermarket Catalogue' (for Student Book activity), Student Book p. 129 'Supermarket Shopping'

TASK 1: PRECISE SPENDING

Give students a toy catalogue and a sheet of A3 paper. They can choose items to buy, spending as close to $100 as they can without going over. Have them aim to be within 10 cents of the goal. Have students cut out selected items and create a poster of the items purchased and total amount spent. Remind students about rounding prices.

TASK 2: INTERACTIVE TASK 

Have students work with a partner on computers to create a toy 'wish list' as an *Excel* spreadsheet. The list should contain at least 20 items from a toy catalogue, with the toy in one column, price in the next and why this toy is 'needed' in the third. Items should be listed in order of price. Once completed, have students estimate the total cost, then use the 'Autosum' function in *Excel* to give the actual cost of the list. Compare the estimate and actual answers.

TASK 3: STUDENT BOOK p. 129 ***'Supermarket Shopping'***

TEACHING GROUP

You will need: toy catalogues, calculators

$50 TO SPEND

- For students who require support, modify the toy catalogue task in Independent Tasks, Task 1, so they only have $50 to spend and use prices that do not need rounding. Give examples of numbers that are more and less than 50.
  Model the process of selecting items, using a calculator to add items as you go, ensuring that the total does not exceed $50. If it does, select a cheaper item instead. Once modelled, students complete their own work. You may provide a more flexible range (e.g. $48–$50) if required.

WHO CAN SPEND THE FASTEST?

- For students who require a challenge, give them each a toy catalogue. Allow them to look through the catalogue so they have an idea of the position of different prices. They are going to race to choose items that total a set amount, e.g. $75. The person to reach the target first (within 5 cents) gets one point. Set a new target, e.g. $40, and play again. To challenge further, select larger numbers involving more calculations. Calculators may be permitted, depending on ability.

REFLECTION

Select from the following to suit your class and their learning outcomes:

- Give students a selection of play money (coins only) or coins made from BLM 87 'Coins'. Have them model various amounts of $5 or less in two different ways.
- Ask, 'Why do we have notes and coins?' Discuss responses.

LESSON PLAN 3

TUNING IN

MAKING CHANGE

Pose the problems: 'I went to the bakery on Monday and bought a loaf of bread. I gave the shopkeeper $2 and got 5 cents change. How much was the bread?' 'I went back on Tuesday and bought a different loaf. The price was $1.99. I gave them $2 but I did not get any change. Why not?' Review the concept of rounding.

WHOLE-CLASS INTRODUCTION

ALL ABOUT CHANGE

You will need: school canteen/lunch order list or BLM 90 'Canteen List', BLM 87 'Coins' (or play money)

Display an enlarged copy of the school canteen list or BLM 90 'Canteen List' on the board. Say, 'I'd like to buy a vegetable pastie and an apple juice for lunch. How much will this cost?' Have students use BLM 87 'Coins' to model their answer. Ask, 'What operation did you use to solve this problem?' Discuss the use of addition. Ask, 'If I gave $5 to pay for my lunch, how much change would I be given?' Discuss responses, and ask, 'What operation do you use now to work this out?' Discuss use of either subtraction or counting up strategies to calculate change. Select another two items from the list and have students work out how much change you would receive from $5. Continue as necessary, varying the items purchased and the amount of money used to pay.

INDEPENDENT TASKS

Note: Choose from Tasks 1, 2 or 3.

You will need: school canteen/lunch order list or BLM 90 'Canteen List', *PowerPoint,* access to the internet, Student Book p. 130 'Calculating Change'

TASK 1: CREATING AND SOLVING CANTEEN PROBLEMS

Have students work in pairs with a copy of the school canteen list or BLM 90 'Canteen List'. Have them create ten problems for their partner using the format: 'If you bought ____ and ____ and paid $____, how much change would you receive?' Students can combine three items from the list in some of their problems to make them more challenging. Students swap with their partner to answer the problems and swap back to correct each other's work.

TASK 2: INTERACTIVE TASK 

Have students work independently on computers to create a *PowerPoint* presentation of their own shop. The first slide has the name of their shop. The next slide contains six items that can be purchased at the shop, including images of the items from the internet. The next three slides contain the purchases of three customers, the total cost, the amount paid and the change given, e.g. Customer 1 bought a shirt, shorts and a pair of socks for $61.50. She paid $70 and received $8.50 in change.

TASK 3: STUDENT BOOK p. 130 *'Calculating Change'*

TEACHING GROUP

You will need: BLM 87 Coins (or play money)

SMALL CHANGE

- For students who require support, discuss the need for change. Ask, 'Do you always have the exact amount to buy things when you go shopping? What happens?' Pose simple change problems, with values of less than $3, e.g. 'I bought two apples from the supermarket that cost 80 cents, I paid with a $2 coin, how much change will I get?' Remind students that $1 is 100 cents. Discuss methods for solving the problem – using subtraction to find the difference, or using coins to count up from 80 cents to $2. Pose other similar problems, e.g. giving $2 for $1.05 or $3 for $2.45, to consolidate understanding.

LIMITING CHANGE

- For students who require a challenge, have them work in pairs to create canteen problems for each other as in Independent Tasks, Task 1, but limit their change to 5-cent, 10-cent and 20-cent pieces. Have them model three different ways the change could be given to each problem.

REFLECTION

Select from the following to suit your class and their learning outcomes:

- Have students show their *PowerPoint* presentations from Independent Tasks, Task 2, to the class, discussing how the purchase cost and change for each of the customers were solved.
- Tell students that up until the early 1990s in Australia, we also had 1-cent and 2-cent coins. Ask, 'Do you think this would have made paying for items and working out change easier or more difficult? Why?'

Home Tasks

Select from the possible Home Tasks:

- Have students join their parents or carers for the weekly grocery shopping trip. Have them first guess how much money the family spends on groceries per week, then keep a rough estimate of the amount spent as items are added to the trolley. Compare the guess, rough estimate and actual amount spent.
- Ask students to bring in any international or Australian 1-cent and 2-cent coins they may have to show the class.

Assessment

- Have students complete **Student Assessment p. 131**.
- Review with students **Assessment Task Card 3.32**.

During the three lessons:

- Collect written reflections from Lesson Plan 1 to show evidence of how students think about their learning.
- Make digital copies of *Excel* spreadsheets and *PowerPoint* presentations to use in digital portfolios and demonstrate use of ICT in the classroom.
- Make note of students who completed their Home Tasks as evidence of the link between home and school.

Recommendations for Future Learning

Specific to Student Assessment p. 131; if the student is experiencing some difficulty:

Q 1–2 Revise the value of Australian coins with the student. Remind the student that 100 cents equals $1. Have the student use play money to model then count the amount.

Q 3 Revisit the 'Using Money' activity from the Whole-Class Introduction in Lesson Plan 2. Review the value of banknotes and coins and the relationships between them, e.g. two $5 notes equals $10 and so on. Use play money to model the amounts if required.

Q 4 Have the student practise using subtraction to work out the difference between the two values, or encourage the use of the counting up strategy, depending on student preference. Review the concept of rounding prices.

If the student has not achieved the recommended skills for this unit:

1. See **Assessment Task Card 3.32** for specific recommendations.
2. Explore 'Counting' from *Nelson Maths Building Mental Strategies Big Book 3*, pp. 2–3.
3. Revise basic strategies for addition and subtraction (difference). Focus on using 10/100 as a base for grouping and adding.
4. Review *Nelson Maths: Australian Curriculum NSW Year 2* Unit 17.

If the student has achieved the recommended skills and these skills are firmly established, consider:

1. Having the student complete *Building Mental Strategies Skill Book Year 4*, pp. 42–43, to further explore the counting-up strategy.
2. Moving forward to *Nelson Maths: Australian Curriculum NSW Year 4* Unit 31.
3. Extending the student by working with more numbers to add, or by using larger numbers.

Year 3: Assessment Task Card

Numbers, Numbers, Numbers
Resources: A4 paper, NTO 3.3 'Six-Sided Dice'

1 Give each student a sheet of paper. Present NTO 3.3 'Six-Sided Dice' with four digits generated on four dice. Have students arrange the digits to form an even number. Have students record the number and read it aloud.

2 Have students use the four digits to form an odd number, record the number and read it aloud.

3 Have students form two other numbers using the four digits and record with the two previous numbers.

4 Have students write all of the numbers in order from smallest to largest.

5 Have students write the numbers in word form.

Whole numbers
MA2-4NA applies place value to order, read and represent numbers of up to five digits

Year 3: Assessment Task Card

Numbers, Numbers, Numbers TARGETED ASSESSMENT

If the student is experiencing difficulty:

Q1–2 Have the student use ten frames and counters to explore odd and even numbers to 10. Each time they determine that a number is odd or even, have them add a ten frame, then another. Discuss with the student that, it does not matter how many tens or hundreds are added to the number, it will remain an odd or even number depending on what is in the ones place.

Q3–4 Begin by having the student locate 2-digit numbers on a number line and use it to compare to other numbers. When the student is able to identify larger or smaller 2-digit numbers by their position on a number line, extend to 3-digit numbers.

Q5 Have the student write the words for the numbers 1 to 9 and then for the numbers 20, 30 up to 90. Have them practise writing in words the numbers 20 to 99. Next have the student practise the numbers 11 to 19. When the student is proficient, have them learn the words for 100 and 1 000 and write the words for 3- and 4-digit numbers.

Whole numbers
MA2-4NA applies place value to order, read and represent numbers of up to five digits

Year 3: Assessment Task Card

Numbers to 10 000
Resources: number cards, MAB, A3 paper

1 Give students the following number cards: 2371, 3612, 6152, 3214, 2534, 6514. Ask them to name each number.

2 Have students select the largest number and model it with MAB.

3 Have students explain how they know it is the largest number.

4 Have students order the numbers from smallest to largest.

5 Ask students to draw a line on a landscape sheet of A3 paper, stick the smallest number at the left-hand end and the largest at the right-hand end. Students stick the remaining numbers on the number line, varying the distance between each to show how much bigger or smaller the numbers are.

Whole numbers
MA2-4NA applies place value to order, read and represent numbers of up to five digits

Year 3: Assessment Task Card

Numbers to 10 000 TARGETED ASSESSMENT

If the student is experiencing difficulty:

Q1 Select one number, cover the thousands digit and have the student name the remaining 3-digit number. Review what a 4-digit number is and reveal the hidden digit. Have the student say this number. Repeat for other numbers.

Q2 Revisit the 'Introducing... Thousands' activity from the Whole-Class Introduction in Lesson Plan 1. Allow the student to model 3-digit numbers with MAB before moving on to 4-digit numbers.

Q3–4 Select two 3-digit numbers and ask the student to model each number with MAB. Discuss with the student that when deciding whether a number is bigger or smaller than another number, we look at the biggest part of the number first. Compare the two models and decide which is biggest. Extend to comparing two 4-digit numbers and gradually to a group of 4-digit numbers.

Q5 Revisit the 'Pegs for Scale!' activity from the Whole-Class Introduction in Lesson Plan 3. Use NTO 3.6 'Number Line' to review location of numbers, comparing larger and smaller. It may be useful to begin with 3-digit numbers and build to 4-digit numbers.

Whole numbers
MA2-4NA applies place value to order, read and represent numbers of up to five digits

Year 3: Assessment Task Card

More About Numbers to 10 000
Resources: A4 paper, spike abacus

1 Give each student a sheet of paper. Have them create six numbers using the digits 2, 4, 6, 8. Ask students to order their numbers from smallest to largest.

2 Have students select the smallest number and model it on a spike abacus.

3 Ask students to discuss the value of each digit in the smallest number.

4 Have students add 10, 100 and 1 000 to the smallest number.

5 Have students subtract 10, 100 and 1 000 from the smallest number.

6 Ask, 'What amount would you need to subtract from your smallest number to have a zero in the hundreds column?'

7 Have students explain the importance of zero in a number.

Whole numbers
MA2-4NA applies place value to order, read and represent numbers of up to five digits

Year 3: Assessment Task Card

More About Numbers to 10 000 TARGETED ASSESSMENT

If the student is experiencing difficulty:

Q1 Revise what a 4-digit number is.

Q2 Allow the student to model the number using MAB as well as the spike abacus.

Q3 Simplify the task by modelling 3-digit numbers using MAB and the place-value chart.

Q4–5 Revisit the 'Adding and Subtracting 10, 100 and 1 000' activity from the Whole-Class Introduction in Lesson Plan 1. Use MAB and the place-value chart to consolidate understanding. It may be useful to begin with 3-digit numbers and build to 4-digit numbers.

Q6–7 Have the student model the number with MAB on a place-value chart. Have the student remove all of the hundreds flats from the chart to get a value of zero in the hundreds. Discuss the importance of writing '0' in this space.

Whole numbers
MA2-4NA applies place value to order, read and represent numbers of up to five digits

Year 3: Assessment Task Card

Length
Resources: A4 paper, 30 cm ruler, metre ruler

1 Give each student a sheet of paper. Have students find three items in the classroom that are about the same length as their foot.
2 Have students estimate the length of their items in centimetres.
3 Have students use a ruler to measure and record the actual length of each object.
4 Have students find three items in the classroom that are about 1 m in length.
5 Have students use a metre ruler to measure the items and say whether they are shorter or longer than a metre.
6 Have students write all six objects in order from longest to shortest.
7 Have students estimate and then measure the distance across the top of a thumb tack.

Length
MA2-9MG measures, records, compares and estimates lengths, distances and perimeters in metres, centimetres and millimetres, and measures, compares and records temperatures

Year 3: Assessment Task Card

Length TARGETED ASSESSMENT

If the student is experiencing difficulty:

Q1 Have the student look at the length of their foot and compare it to different items in the room.
Q2 Ask the student to look at the size of a centimetre and think about how long a 30 cm ruler is. Have the student use this knowledge to predict the length of the selected items.
Q3 Revisit activities from Lesson Plan 1 where centimetres are used to measure items accurately.
Q4 Give the student a metre ruler (or 1 m piece of string), which they can use to find objects of about 1 m.
Q5 Have the student hold a metre ruler against the object and decide whether the object is shorter or longer than 1 m.
Q6 Remind the student of the meanings of the words 'shortest' and 'longest'. Revise ordering activities from Unit 2 Numbers to 10 000, e.g. Lesson Plan 2, Independent Tasks, Task 2.
Q7 Ensure students are beginning measuring at the 'zero point' on the ruler.

Length
MA2-9MG measures, records, compares and estimates lengths, distances and perimeters in metres, centimetres and millimetres, and measures, compares and records temperatures

Year 3: Assessment Task Card

Mental Strategies for Addition

Resources: A4 paper, NTO 3.10 'Card Flip', dice, BLM 15 'Number Cards 1–20'

1 Give each student a sheet of paper. Present NTO 3.10 'Card Flip', selecting 1–100 cards. Have students add 10, 60 and 100 to each number shown. Ask them to explain the strategy they used.

2 Present NTO 3.10 'Card Flip', selecting 1–100 cards. As the numbers are shown, have students roll a dice and record an addition equation, e.g. 45 + 6. Students solve the equations and explain the strategy they used.

3 Have students select a card made from BLM 15 'Number Cards 1–20'. Ask them to write a doubles equation and a near doubles equation that uses their number.

4 Have students record a list of the mental strategies they can use when solving addition equations.

Addition and subtraction

MA2-5NA uses mental and written strategies for addition and subtraction involving two-, three-, four- and five-digit numbers

Year 3: Assessment Task Card

Mental Strategies for Addition

TARGETED ASSESSMENT

If the student is experiencing difficulty:

Q1 Provide opportunities for the student to explore adding multiples of 10, e.g. 40 + 20. Support the student in modelling the equations using MAB or NTO 3.7 'Modelling with MAB'.

Q2 Revise the numbers that add together to make 10. Show the student numbers using NTO 3.10 'Card Flip' (selecting 1–100 cards) and have the student identify how many more is needed to build to the next ten from that number. When the student is proficient, encourage them to apply this strategy when solving equations such as 25 + 9.

Q3 Revise what doubles equations are. Have the student use counters to model how the numbers are equal. Revise what near doubles equations are and have the student use counters to model how one number is bigger. Discuss how being familiar with doubles can help you solve near doubles.

Q4 Discuss what strategies are and how they help us with solving addition problems. Encourage the student to 'talk aloud' as they solve problems to identify what they are thinking and doing as they solve problems.

Addition and subtraction

MA2-5NA uses mental and written strategies for addition and subtraction involving two-, three-, four- and five-digit numbers

Year 3: Assessment Task Card

Addition
Resources: A4 paper

1 Give each student a sheet of paper. Present the equations: 46 + 28 and 53 + 39. Have students write the equations vertically and record the answers.

2 Have students estimate the answers to the equations by rounding, then check the reasonableness of their answers.

3 Have students generate their own 2-digit with 2-digit addition equations and find the answers.

4 Present the problem: 26 students in 3C and 28 students in 3W were going on an excursion to the zoo. How many students were going? Have the students solve the problem.

Addition and subtraction
MA2-5NA uses mental and written strategies for addition and subtraction involving two-, three-, four- and five-digit numbers

Year 3: Assessment Task Card

3.6

Addition TARGETED ASSESSMENT

If the student is experiencing difficulty:

Q1 Have the student solve 2-digit with 2-digit addition problems by modelling the numbers with MAB using NTO 3.7 'Modelling with MAB'. Model how the ones are added together and the tens are added together. Review how the ones are regrouped to form a ten when the total of ones is greater than 9.

Q2 Review the rules about rounding numbers to the nearest ten. Use NTO 3.10 'Card Flip', selecting 1–100 cards, to present the student with a 2-digit number. Have the student round the number up or down. Also revise adding multiples of 10, using MAB to model if necessary. Present students with two 2-digit numbers and have them round both and then add the multiples of 10 together.

Q3 Give the student a set of cards made from BLM 5 'Digit Cards'. Have them rearrange the cards to form familiar addition equations, i.e. have cards representing the addends and cards representing the answer. Begin with 1-digit addition equations and progress to 2-digit equations. Discuss with the student how they need to consider the ones being added together, the tens being added together and regrouping that may occur.

Q4 Discuss what the problem is asking. Have the student identify the numbers (the addends) in the problem. Talk about specific strategies the student could use to solve the problem.

Addition and subtraction
MA2-5NA uses mental and written strategies for addition and subtraction involving two-, three-, four- and five-digit numbers

Year 3: Assessment Task Card

Position

Resources: maps of the school, BLM 10 'Grid Paper'

1 Give each student a map of the school. Have them label the different buildings and areas on the map.

2 Have students choose two positions on the map and mark them with a cross, e.g. their classroom and the art room. Have them draw the pathway they would take to move between the two positions.

3 Have students write the directions they would give to someone who did not know how to get from their classroom to the basketball courts.

4 Give students BLM 10 'Grid Paper'. Have them draw a map of a familiar place, e.g. their bedroom, house or classroom. Have them label the different positions and items on their map.

5 Have students write coordinates for different positions and items on their map.

Position

MA2-17MG uses simple maps and grids to represent position and follow routes, including using compass directions

Year 3: Assessment Task Card

Position

TARGETED ASSESSMENT

If the student is experiencing difficulty:

Q1–2 Revise how to read and interpret maps. Discuss how a map is a bird's-eye view of a position. Discuss how we can use maps to help locate places and to help us make a pathway between two places. Have the student practise locating specific positions on a map. Give them opportunities to create and follow pathways between two positions on a map.

Q3 Review specific language that can be used for giving directions. Ensure the student understands the meaning of words such as 'forwards', 'backwards', 'left' and 'right'. Give the student simple directions to follow around the room. Then have the student give directions for others to follow.

Q4 Revise specific strategies for drawing maps. Discuss how maps need to be drawn from a bird's-eye view. Talk about how they need to consider the size, shape and position of items that are on a map.

Q5 Review grid maps and coordinates. Revise how coordinates are written and used. Have the student find specific positions and coordinates on grid maps. Support the student in using coordinates to describe specific positions and pathways on their map.

Position

MA2-17MG uses simple maps and grids to represent position and follow routes, including using compass directions

Year 3: Assessment Task Card

Place Value
Resources: A3 paper, MAB

1 Give each student a sheet of A3 paper. Have them fold it into quarters and open it again to form four sections.

2 Have students write a 4-digit number between 2 000 and 3 000 in the centre of the page (where the four corners meet).

3 Model this number using MAB.

4 In one section, students write their 4-digit number in words.

5 In another section, students expand and rename their number using no hundreds.

6 In another section, students partition their number.

7 In the last section, students add 300 to their number.

Whole numbers
MA2-4NA applies place value to order, read and represent numbers of up to five digits

Year 3: Assessment Task Card

Place Value TARGETED ASSESSMENT

If the student is experiencing difficulty:

Q2–3 Review what a 4-digit number is. Revisit the place-value chart and discuss the name of each column. Have the student say the number aloud before writing it.

Q4 Remind the student of the value of each MAB. It may be useful to set them out on a place-value mat.

Q5 Revisit the 'Oh No! No Hundreds!' activity from the Whole-Class Introduction in Lesson Plan 1. Use BLM 23 '4-Digit Number Expanders' to assist students.

Q6 Revisit the 'Partition' activity from the Whole-Class Introduction in Lesson Plan 2. Model numbers with MAB and link this to partitioning.

Q7 Simplify the task by beginning with a 3-digit number and adding 100, then 200 and then 300.

Whole numbers
MA2-4NA applies place value to order, read and represent numbers of up to five digits

Year 3: Assessment Task Card

More About Place Value
Resources: A4 paper, MAB, dice

1 Give each student a sheet of paper. Have them write their code for ones, tens, hundreds and thousands across the top of the page and draw a box around it.

2 Have students write a 4-digit number between 3 000 and 4 000 below their code.

3 Have students write the number in code and model it with MAB.

4 Have students roll a dice and add that number of hundreds to their number. Model with MAB and record the number.

5 Have students roll the dice and subtract that number of tens from their number from Q4. Model with MAB and record the number.

6 Have students decide whether their number from Q2 or Q5 is closer to 3 500.

Whole numbers
MA2-4NA applies place value to order, read and represent numbers of up to five digits

Year 3: Assessment Task Card

More About Place Value TARGETED ASSESSMENT

If the student is experiencing difficulty:

Q1 Revisit the 'A Class Code' activity from the Whole-Class Introduction in Lesson Plan 1.

Q2 Review what a 4-digit number is. Use the place-value mat and MAB if necessary.

Q3 Remind the student of the value of each symbol in their code. It may be useful to set out the MAB on a place-value mat.

Q4–5 Encourage the student to model their number before answering and manually add or subtract the amount on the dice. If trading is necessary, remind students that 10 tens make 1 hundred and so on.

Q6 Discuss whether numbers are bigger or smaller than 3 500. Use MAB to model both and show what is needed to bring each to 3 500. Revise the domino target activities from Lesson Plan 3.

Whole numbers
MA2-4NA applies place value to order, read and represent numbers of up to five digits

Year 3: Assessment Task Card

Mass
Resources: A4 paper, a collection of classroom items, beam balances, kitchen scales

1. Give each student a sheet of paper and a collection of classroom items. Ask them to predict the order of items from lightest to heaviest.
2. Have students select measuring devices to find the mass of each item and record their findings.
3. Have students order the items from lightest to heaviest according to their actual weight.
4. Have students compare their predictions to the actual order and explain any differences.
5. Have students find and list three classroom items that are lighter than their lightest item.
6. Have students find and list three classroom items that are heavier than their heaviest item.

Mass
MA2-12MG measures, records, compares and estimates the masses of objects using kilograms and grams

Year 3: Assessment Task Card

Mass TARGETED ASSESSMENT

If the student is experiencing difficulty:

Q1 Have the student look at and feel the items to decide the order from lightest to heaviest, comparing two items at a time.

Q2 Remind the student of the class activities for using both the beam balance and scales and what they were used for – it may be easier to weigh heavier items on the scales.

Q3 Have the student look at the value of their numbers to order them. You may use a place-value mat to write the numbers and/or MAB to model the numbers.

Q4 Have the student look for any differences and apply reasoning as to why their results are different.

Q5-6 Remind the student of the meaning of 'lighter' and 'heavier'. Have them use informal methods of comparison and/or place two items on a beam balance to check.

Mass
MA2-12MG measures, records, compares and estimates the masses of objects using kilograms and grams

Year 3: Assessment Task Card

Our Community – Data
Resources: A4 paper, graph paper

1 Give each student a sheet of paper. Have students think of four different research questions that they could ask their friends.
2 Ask students to select one question to ask their friends and develop at least five response categories.
3 Have students create a data collection table, then collect responses from ten friends.
4 Once students have collected their data, have them make a graph of their results on graph paper.
5 Have students write three facts that are shown by their graph.

Data
MA2-18SP selects appropriate methods to collect data, and constructs, compares, interprets and evaluates data displays, including tables, picture graphs and column graphs

Year 3: Assessment Task Card

3.11

Our Community – Data

TARGETED ASSESSMENT

If the student is experiencing difficulty:

Q1 Discuss with the student the types of questions that make good research questions for graphing, i.e. those that can be divided into five or six response categories. If needed, brainstorm a small list of possible questions on the board, starting with 'What is your favourite [colour, animal, milkshake flavour etc.]?' Have students select one to work with.

Q2 Have the student think of the most common responses to their question, and list five or six. Remind the student that the categories cannot overlap and if there are a few main categories but lots of other options, they may use an 'other' category.

Q3 Give the student BLM 29 'Data Collection Table' if they are having trouble developing their own.

Q4 Give the student BLM 30 'Graph It!' to help them complete their graph.

Q5 Revise the work done in the interpreting data activities in Lesson Plan 3, reminding the student that their findings need to be facts that are shown in the graph. Have the student explain what the graph is showing to ensure they have a clear understanding. Encourage comparison between columns using 'more' or 'less' statements.

Data
MA2-18SP selects appropriate methods to collect data, and constructs, compares, interprets and evaluates data displays, including tables, picture graphs and column graphs

Year 3: Assessment Task Card

Mental Strategies for Subtraction
Resources: A4 paper, NTO 3.10 'Card Flip'

1 Give each student a sheet of paper. Have students solve the following subtraction equations:
14 – 5   17 – 8   11 – 4   24 – 6

2 Have students explain the strategy they used to solve the subtraction equations.

3 Present the equation: 15 – 3 = 12. Have students solve the following problems and explain the strategy they used.
25 – 3   35 – 3   45 – 3   55 – 3

4 Present the equation: 12 – 5 = 7. Have students write other problems this equation could help them solve.

5 Present NTO 3.10 'Card Flip', selecting 1–100 cards. Have students subtract 10, 30 or 50 from the number shown. Ask them to explain the strategy they used.

Addition and subtraction
MA2-5NA uses mental and written strategies for addition and subtraction involving two-, three-, four- and five-digit numbers

Year 3: Assessment Task Card

Mental Strategies for Subtraction — TARGETED ASSESSMENT

If the student is experiencing difficulty:

Q1–2 Have the student revise strategies for solving subtraction problems, e.g. taking away, counting back, counting on, finding the difference. Provide opportunities for the student to solve subtraction problems using the strategies.

Q3–4 Have the student revise serial subtraction. Present the equations: 17 – 5 = 12, 27 – 5 = 22. Have the student discuss what they notice. Present the equations: 37 – 5, 47 – 5, 57 – 5. Discuss strategies for solving these equations. Use number lines and a 100 chart for students to explore other serial subtraction patterns.

Q5 Have the student subtract multiples of 10, e.g. 70 – 40. Encourage the student to model the total number with MAB and then take away the appropriate amount. When the student is proficient, move on to equations such as 76 – 40. Continue to have the student model the equation with MAB. Emphasise that when tens are taken away, the ones remain the same and the tens decrease.

Addition and subtraction
MA2-5NA uses mental and written strategies for addition and subtraction involving two-, three-, four- and five-digit numbers

Year 3: Assessment Task Card

Subtraction
Resources: A4 paper

1 Give each student a sheet of paper. Present the equations: 59 – 23, 62 – 38. Have students write the equations vertically and record the answers.

2 Have students estimate the answers to the equations by rounding to the nearest ten. Have them check the reasonableness of their answers to the equations.

3 Have students generate their own 2-digit subtraction equations and find the answers. Ask, 'Can you make a subtraction problem that involves renaming?'

4 Present the problem: 'There were 45 students playing on the playground. 16 ran off to play on the oval. How many students were left on the playground? Have students solve the problem.

Addition and subtraction
MA2-5NA uses mental and written strategies for addition and subtraction involving two-, three-, four- and five-digit numbers

Year 3: Assessment Task Card

Subtraction TARGETED ASSESSMENT

If the student is experiencing difficulty:

Q1 Have the student solve 2-digit subtraction problems by modelling the numbers with MAB using NTO 3.7 'Modelling with MAB'. Model how we subtract the ones and then subtract the tens. Review how a ten can be renamed for 10 ones if there are not enough ones.

Q2 Review the rules about rounding numbers to the nearest ten. Use NTO 3.10 'Card Flip' (select 1–100 cards) to present the student with a 2-digit number. Have the student round the number up or down. Also revise subtracting multiples of 10, using MAB to model if necessary. Present the student with two 2-digit numbers and have them form a subtraction problem using these numbers. Have the student estimate the answer by rounding both of the numbers to the nearest ten and then subtracting them.

Q3 Give the student a set of cards made from BLM 5 'Digit Cards'. Have them the rearrange cards to form familiar subtraction equations. Begin by forming 1-digit subtraction equations and progress to forming 2-digit equations. Remind the student that they need to consider how the ones are subtracted, the tens are subtracted and also any renaming that may need to occur.

Q4 Have the student discuss what the problem is asking. Have the student identify the numbers in the problem and discuss what the numbers mean in relation to the problem. Talk about specific strategies the student could use to solve the problem.

Addition and subtraction
MA2-5NA uses mental and written strategies for addition and subtraction involving two-, three-, four- and five-digit numbers

Year 3: Assessment Task Card

Connections Between Addition and Subtraction
Resources: A4 paper

1 Give each student a sheet of paper. Present students with the numbers 8, 6 and 14. Have them record the addition and subtraction problems within this fact family.

2 Present students with the following addition and subtraction problems. Have them solve the problems and then record a related problem that proves their answer is correct.

26 + 35 73 – 46 39 + 27

3 Have students solve equations with missing numbers:

25 + ____ = 37 – 4 46 – 13 = 24 + ____ 9 + 3 + 7 = 25 – ____

4 Give students an equation such as 45 + 13 = 58. Ask them to list other equations this might help them solve.

Addition and subtraction
MA2-5NA uses mental and written strategies for addition and subtraction involving two-, three-, four- and five-digit numbers

Year 3: Assessment Task Card

Connections Between Addition and Subtraction TARGETED ASSESSMENT

If the student is experiencing difficulty:

Q1 Have the student model the number 8 with red counters and 6 with blue counters. Establish that when these numbers are moved together there is a total of 14 counters. Take away six or eight of the counters and have the student predict how many are left. Discuss how the student knew what the answer would be. Record the addition and subtraction problems.

Q2 Have the student solve the addition problem 15 + 3, then have them solve 18 – 3. Discuss what the student has noticed with these two problems. Discuss how these problems are related because they involve the same numbers.

Q3 Revise what balanced equations are. Draw an equals symbol in the middle of the board and have the student record equations on each side of the symbol that equal the same amount. Discuss strategies for working out the missing number, e.g. solving one side of the equation and using this answer to help solve the other side.

Q4 Discuss how addition and subtraction problems are connected. Have the student rearrange the numbers in the problems to make other addition and subtraction problems. Use NTO 3.7 'Modelling with MAB' to model the problems and show the connection between these numbers.

Addition and subtraction
MA2-5NA uses mental and written strategies for addition and subtraction involving two-, three-, four- and five-digit numbers

Year 3: Assessment Task Card

Solving Addition and Subtraction Problems

Resources: A4 paper

1 Give each student a sheet of paper. Have students solve the following addition and subtraction problems.

425 + 627   549 + 375   783 – 249

634 – 255   531 – 174   328 + 539

2 Present students with the following addition and subtraction problems. Have them write a worded problem to represent each.

236 + 135   743 – 426

3 Ask students to select two of the following numbers: 785, 846, 254, 127, 436. Students choose to add or subtract the two numbers, with the aim of getting as close as they can to the target number: 653.

4 Ask, 'How do you know that the two numbers you have chosen get you closet to the target number?'

Addition and subtraction

MA2-5NA uses mental and written strategies for addition and subtraction involving two-, three-, four- and five-digit numbers

Year 3: Assessment Task Card

Solving Addition and Subtraction Problems

TARGETED ASSESSMENT

If the student is experiencing difficulty:

Q1 Have the student solve addition and subtraction problems with 2-digit numbers before moving on to problems involving 3-digit numbers. Have the student use MAB to model the equations. Revise how to rename or regroup hundreds, tens and ones in addition and subtraction problems.

Q2 Have the student use NTO 3.7 'Modelling with MAB' to model addition and subtraction problems. Discuss what the student notices about the problems. Identify and discuss addition and subtraction problems in real-life situations. Have the student connect given problems to real-life situations.

Q3–4 Change the numbers the student can choose from to 2-digit numbers and set a 2-digit target number. Revise the difference between addition and subtraction problems. Discuss how the student could reach the target by adding two smaller numbers together or subtracting from a larger number.

Addition and subtraction

MA2-5NA uses mental and written strategies for addition and subtraction involving two-, three-, four- and five-digit numbers

Year 3: Assessment Task Card

Time

Resources: six clock faces from BLM 45 'Blank Clock Faces', A4 paper, glue

1 Give each student BLM 45 'Blank Clock Faces' and a sheet of paper. Have them show the following times on six of their clock faces:

- the time they have lunch
- the time they get up in the morning
- the time they get home from school
- the time they eat breakfast
- the time they go to bed
- the time their favourite TV show starts.

2 Have students cut out and order the six clock faces from earliest to latest and glue them to their sheet of paper.

3 Have students write the digital time underneath each of the clock faces.

4 Have students write three things they have to remember when reading an analogue clock.

Time

MA2-13MG reads and records time in one-minute intervals and converts between hours, minutes and seconds

Year 3: Assessment Task Card

Time

TARGETED ASSESSMENT

If the student is experiencing difficulty:

Q1 Revise how to show time on an analogue clock and how to use the hour and minute hands. Have the student think about one hand at a time. Revise the strategy of counting by 5s to find minutes.

Q2 Have the student think logically about which activity would come first in their day, and use this practical knowledge to order their clocks.

Q3 Remind the student of the format used in digital time. Show examples of digital times on NTO 3.18 'Clocks' or on digital devices in the classroom.

Q4 Encourage the student to think about the two hands on the clock, the numbers 1–12 and what they mean, how many minutes there are in one hour, and so on.

Time

MA2-13MG reads and records time in one-minute intervals and converts between hours, minutes and seconds

Year 3: Assessment Task Card

Angles

Resources: A4 paper, clock with movable hands

1. Give each student a sheet of paper and have them draw the following lines:
   - vertical
   - curved
   - horizontal
   - diagonal.
2. Have students give an example of where they have seen each of these lines.
3. Have students explain how angles are formed.
4. Have students identify examples of angles around the room that are less than, more than or exactly a right angle.
5. Have students use a clock with movable hands to demonstrate how an angle equal to a quarter turn is formed.
6. Have students give an example of angles that are equal to a quarter turn.

Angles

MA2-16MG identifies, describes, compares and classifies angles

Year 3: Assessment Task Card

Angles

TARGETED ASSESSMENT

If the student is experiencing difficulty:

Q1 Review the different types of lines: curved, diagonal, horizontal and vertical. Have the student describe how the lines are similar and different.

Q2 Have the student practise drawing each type of line. Have the student identify the four types of lines in familiar objects. Discuss how objects can include a combination of different lines.

Q3–4 Revise what an angle is. Have the student explore how angles are formed when two lines intersect. Encourage students to find different angles and to use a right-angle tester to compare the angles to a right angle. Discuss how angles are also formed through rotation of lines. Have the student move the hands of a clock to form angles of different sizes.

Q5–6 Revise the concept of a right angle as a quarter of a turn. Have the student rotate one hand on a clock a quarter turn and describe the angle that is formed. Have them then find angles in the room that are the same size as the angle shown on the clock.

Angles

MA2-16MG identifies, describes, compares and classifies angles

Year 3: Assessment Task Card

Mental Strategies for Multiplication
Resources: A4 paper, 10-sided dice

1 Have students skip count by 2s up to 50. Ask them how this could help them multiply numbers by 2.

2 Give each student a sheet of paper. Present the following equations for students to solve. Have them explain the strategy they used.

5 × 2 8 × 2 6 × 2 10 × 2

3 Have the students roll a 10-sided dice four times and multiply each number rolled by 1. Have them explain how they use place value to solve the problems.

4 Have the students roll the dice another four times and multiply each number rolled by 10. Have them explain the strategy they used to solve the problems.

Multiplication and division
MA2-6NA uses mental and informal written strategies for multiplication and division

Year 3: Assessment Task Card

3.18

Mental Strategies for Multiplication TARGETED ASSESSMENT

If the student is experiencing difficulty:

Q1 Have the student use counters to make groups of 2. Have them skip count the groups and record these numbers. Have the student use this to help them record 2 times table facts.

Q 2 Have the student use counters to model the multiplication facts. Emphasise that 2 times table facts are made of groups of 2. Then have the student double the numbers. Guide the student to recognise that doubling a number gives the same answer as forming that many groups of 2.

Q 3–4 Have the student model ×1 and ×10 equations using MAB ones and tens. Emphasise that 1 times table facts can be modelled using ones because they are groups of 1. Emphasise that 10 times table facts can be modelled using tens because they are groups of 10.

Multiplication and division
MA2-6NA uses mental and informal written strategies for multiplication and division

Year 3: Assessment Task Card

3.19

Unit 19

More About Mental Strategies for Multiplication
Resources: A4 paper, NTO 3.17 'Ten-Sided Dice'

1 Give each student a sheet of paper. Present them with NTO 3.17 'Ten-Sided 'Dice'. Roll the dice three times and have students multiply each number rolled by 5. Have students explain the mental strategy they used.

2 Roll the dice three more times and have students multiply the numbers rolled by 3. Have them explain the strategy they used.

3 Present the following numbers: 15, 20, 8, 60. Explain that these numbers are the answer to multiplication problems. Have them think of a multiplication problem for each number and explain the strategy they used.

4 Have students solve these problems and explain how they solved them quickly and accurately.

7 × 2 9 × 10 8 × 1 3 × 5 8 × 3

Multiplication and division
MA2-6NA uses mental and informal written strategies for multiplication and division

Year 3: Assessment Task Card

3.19

Unit 19

More About Mental Strategies for Multiplication TARGETED ASSESSMENT

If the student is experiencing difficulty:

Q1 Review the strategy for multiplying numbers by 5, i.e. multiplying by 10 and halving. Have the student use counters to model multiplying numbers by 5, i.e. making groups of 5. Then have them multiply the same numbers by 10 and halve the answer by modelling with MAB. Emphasise that they have the same result.

Q2 Revise the number sequence of counting by 3s. Discuss how being familiar with these numbers helps with multiplying numbers by 3. Have the student solve 3 times table facts by identifying the answers in the counting by 3s number sequence.

Q3 Have the student revise the mental strategies for multiplication with which they are familiar. Have them develop multiplication problems using the strategies.

Q4 Review the mental strategies for multiplication that students have explored. Encourage the student to model the strategies.

Multiplication and division
MA2-6NA uses mental and informal written strategies for multiplication and division

Year 3: Assessment Task Card

Chance

Resources: NTO 3.22 'Spinner' or one of the spinners from BLM 53 'Spinners', A4 paper, blue and yellow counters, paper bag

1 Present NTO 3.22 'Spinner' or one of the spinners from BLM 53 'Spinners'. Ask, 'If I want to have a spinner that can have outcomes of yellow and red segments, but where yellow is more likely to occur, what would it look like?'

2 Give each student a sheet of paper. Have students test their spinner 20 times and record their results on the table below. (Students use either NTO 3.22 'Spinner' or one of the spinners from BLM 53 'Spinners'.)

Yellow	Red

Ask, 'Were the results what you expected? Explain.'

3 Have students place a combination of seven blue and yellow counters in a paper bag so that they are likely to get results shown in the table below, when drawing out a counter and replacing it 21 times.

Blue	Yellow
12	9

Chance

MA2-19SP describes and compares chance events in social and experimental contexts

Year 3: Assessment Task Card

3.20

Chance TARGETED ASSESSMENT

If the student is experiencing difficulty:

Q1 Provide opportunities for the student to explore spinners, using LO: *L2378 'Spinners: predict and test'*, LO: *L2379 'Spinners: spin and label'* and LO: *L2380 'Spinners: explore'*.

Q2 Have the student conduct further chance experiments such as flipping coins, rolling dice and drawing playing cards. Have them work out the possible outcomes and identify the most likely outcome/s, then test to see if results match predictions.

Q3 Give the student probability statements such as '3 out of 4 chance of a result of blue', 'an even chance of white and black' and 'unlikely to result in purple, but likely to result in yellow'. Have the student colour spinners or place counters in a bag to match the statements.

Chance

MA2-19SP describes and compares chance events in social and experimental contexts

Year 3: Assessment Task Card

Patterns
Resources: A4 paper

1 Give each student a sheet of paper. Give them a number pattern, e.g. 4, 7, 10, 13, 16, and have them:
- write the next three numbers in the sequence
- write the rule for the pattern.

2 Have students develop 1-step number patterns using the following rules:
- starting at 4, add 5
- starting at 40, subtract 4.

3 Have students develop 2-step number patterns using the following rules:
- starting at 20, add 7 and subtract 3
- starting at 2, double and add 1.

4 Have students create a 2-step rule for a number pattern that they think is challenging and write the first ten numbers in the pattern.

Patterns and algebra
MA2-8NA generalises properties of odd and even numbers, generates number patterns, and completes simple number sentences by calculating missing values

Year 3: Assessment Task Card

Patterns

TARGETED ASSESSMENT

If the student is experiencing difficulty:

Q1 Revise what a number pattern is. Model the numbers on a 100 chart or number line, look for a pattern and use counting to find the next numbers in the sequence.

Q2 Revise what a rule is and how we use them in number patterns. Place the numbers on a number line to show the progression of the sequence.

Q3–4 Review the work on 2-step rules from Lesson Plan 3. Simplify the rules to start at zero and contain only addition.

Patterns and algebra
MA2-8NA generalises properties of odd and even numbers, generates number patterns, and completes simple number sentences by calculating missing values

Year 3: Assessment Task Card

Multiplication
Resources: A4 paper, dice

1. Give each student a sheet of paper. Have students roll two dice and use the numbers rolled to draw an array.
2. Have students write the multiplication problem to match their array.
3. Have students write a worded problem to represent their multiplication equation.
4. Have students solve the following problems:

$$\begin{array}{r} 23 \\ \times\ \underline{6} \end{array} \qquad 16 \times 2 \qquad \begin{array}{r} 33 \\ \times\ \underline{4} \end{array} \qquad 13 \times 5$$

Multiplication and division
MA2-6NA uses mental and informal written strategies for multiplication and division

Year 3: Assessment Task Card

Multiplication

TARGETED ASSESSMENT

If the student is experiencing difficulty:

Q1–2 Review multiplication problems and arrays. Present arrays to the student using counters and have them record the multiplication problem by counting the rows and the number of counters in each row. Have the student model their own arrays using the counters.

Q3 Provide opportunities for the student to represent multiplication problems using materials. Have them write worded problems to match the multiplication problems.

Q4 Review the process of multiplying 2-digit by 1-digit numbers. Have the student use NTO 3.7 'Modelling with MAB' to model the multiplication problem. Emphasise the process of multiplying the ones and then multiplying the tens. Revise the process of regrouping 10 ones as 1 ten where necessary.

Multiplication and division
MA2-6NA uses mental and informal written strategies for multiplication and division

Year 3: Assessment Task Card

Area

Resources: A4 paper, different sized and shaped sheets of paper and/or a collection of classroom items to measure (e.g. task book, books, pencil cases) and measure with (e.g. playing cards, sticky notes, counters, BLM 10 'Grid Paper')

1. Give students a sheet of paper and a collection of four objects to measure or a prepared collection of different sized and shaped pieces of paper. Ask students to order the objects from those with the smallest area to those with the greatest area. Have students explain their reasoning.
2. Ask students to select two measuring units from a range of uniform units, either formal or informal. Have students explain their choice.
3. Ask students to estimate the area of each object using both measuring units and record their estimates.
4. Ask students to find the actual area of each object using both measuring units and record their results.
5. Have students compare their results with their estimates. Discuss how accurate they were and why.
6. Have students compare their results with other students who used the same units. Are they the same/different? Why?

Area

MA2-10MG measures, records, compares and estimates areas using square centimetres and square metres

Year 3: Assessment Task Card

Area

TARGETED ASSESSMENT

If the student is experiencing difficulty:

Q1 Remind the student that area is the size of a surface. Have them measure by eye to decide which has the smallest area and so on.

Q2 Remind the student of the various informal measures used in Lesson Plan 1 and the square centimetre in Lesson Plan 2. Once chosen, encourage students to explain why they made their choices.

Q3 Remind the student that an estimate is a guess. Have them look at the object and at their measuring device and decide how many they think will fit on their object.

Q4 Revisit activities from Lesson Plans 1 and 2 where students use informal measures and square centimetres to measure items.

Q5 Have the student look at their estimate and their actual measurement. Are they close? Why?

Q6 If results are different, have the student compare the way they measured the area and think about why this could lead to varying results.

Area

MA2-10MG measures, records, compares and estimates areas using square centimetres and square metres

Year 3: Assessment Task Card

Division
Resources: A4 paper

1 Give each student a sheet of paper. Have them solve the following division problems:

18 ÷ 3   32 ÷ 8   27 ÷ 4   45 ÷ 5

2 Have students write a division story to represent the following problem: 21 ÷ 3.

3 Have students explain how they would solve the following problem:

A cake cost $36 and some friends each put in $6 dollars to buy it. How many people were buying the cake?

4 Have students solve these division problems that have remainders:

35 ÷ 8   24 ÷ 5   16 ÷ 3   51 ÷ 7

5 Have students explain how they would solve the following problem: 432 ÷ 3.

Multiplication and division
MA2-6NA uses mental and informal written strategies for multiplication and division

Year 3: Assessment Task Card

Division TARGETED ASSESSMENT

If the student is experiencing difficulty:

Q1 Review strategies for solving division problems. Have the student use counters or MAB to model sharing the total amount into groups. Revise how a ten can be renamed as 10 ones so that it can be shared among the groups.

Q2–3 Revise how some division problems can be solved by calculating how many groups can be made out of a number. Discuss how some division problems tell us how many are in each group, which means we need to find out the number of groups. Use NTO 3.6 'Number Line' to solve division problems by 'jumping back' the number in each group and counting the number of 'jumps' to identify the answer.

Q4 Revise how some division problems involve numbers that cannot be shared evenly among the groups. Discuss how the leftovers are called remainders. Have the student solve division problems using counters or MAB and identify the remainders.

Q5 Begin with the student solving division problems involving 2-digit numbers. Have them use MAB to model renaming a ten as 10 ones. When they are proficient, introduce division with 3-digit numbers. Have the student explore how a hundred can be renamed as 10 tens, as well as a ten being renamed as 10 ones.

Multiplication and division
MA2-6NA uses mental and informal written strategies for multiplication and division

Year 3: Assessment Task Card

More About Division
Resources: A4 paper, counters

1 Give each student a sheet of paper. Have them use counters to make the following arrays:
- four rows of 2
- five rows of 3.

2 Have students record a division problem to match each array.

3 Present students with the following division problems and have them record the multiplication fact they would use to help solve each one:

30 ÷ 3   15 ÷ 5   16 ÷ 2

4 Present students with the numbers 2, 5 and 10. Have them record the multiplication and division problems within this fact family.

Multiplication and division
MA2-6NA uses mental and informal written strategies for multiplication and division

Year 3: Assessment Task Card

More About Division — TARGETED ASSESSMENT

If the student is experiencing difficulty:

Q1 Have the student model arrays and explore how they relate to division problems. Have the student look at the array and identify the total amount, the number of rows and the number in each row. Have them record the equation.

Q2–3 Revise the connection between multiplication and division. Use materials such as counters to model the connection between a multiplication fact and a division fact. Discuss how being familiar with a multiplication fact can help solve a division fact. Have the student record multiplication facts that could help solve division problems.

Q4 Have the student use counters to model the multiplication problems and division problems. Model the equations $2 \times 5 = 10$ and $10 \div 2 = 5$. Discuss what the student has noticed. Emphasise that the equations involve the same numbers. Have the student rearrange the counters to form the other multiplication and division problems.

Multiplication and division
MA2-6NA uses mental and informal written strategies for multiplication and division

Year 3: Assessment Task Card

3.26

Unit 26

Fractions

Resources: BLM 68 'Fraction Cards', BLM 10 'Grid Paper', A4 paper, glue

1. Give each student five fraction cards made from BLM 68 'Fraction Cards' (these may be random or selected depending on ability), a copy of BLM 10 'Grid Paper' and a sheet of paper.
2. Ask students to name each of their fractions.
3. Have students use the grid paper to draw each fraction as a partitioned square or rectangle and shade the appropriate number of squares to represent the fraction.
4. Have students divide their A4 paper into three sections – 'close to zero', 'close to half' and 'close to 1' – then sort and paste the fractions into their appropriate category.
5. Have students select two of their fraction drawings and write a sentence about how they knew which category each drawing belonged to.

Fractions and decimals

MA2-7NA represents, models and compares commonly used fractions and decimals

Year 3: Assessment Task Card

3.26

Unit 26

Fractions

TARGETED ASSESSMENT

If the student is experiencing difficulty:

Q2 Revise what a fraction is and how we say them, e.g. $\frac{2}{5}$ is 'two fifths'.

Q3 Revise the role of the top number (the numerator) and bottom number (the denominator) in fractions. The denominator tells us how many equal parts we are dividing our shape into (number of squares) and the top number tells us how many of the parts we are talking about (shading).

Q4–5 Review the work on fraction size from Lesson Plan 3. Have the student fold their grid paper shape in half and look at whether the shaded section is close to the folded line (close to half), a lot smaller than the line (close to zero) or if the shape is almost fully shaded (close to 1). Use this knowledge to construct sentences.

Fractions and decimals

MA2-7NA represents, models and compares commonly used fractions and decimals

Year 3: Assessment Task Card

More About Fractions
Resources: fraction wall displayed in classroom, A4 paper, counters

1 Give each student a sheet of paper and have them write the following fractions at the top of their page: $\frac{1}{5}, \frac{1}{3}, \frac{1}{2}, \frac{1}{8}, \frac{1}{4}$.

2 Have students order the fractions from smallest to biggest.

3 Using the fraction wall, have students complete the following sentences:

- ___ is the same as ___
- ___ is bigger than ___
- ___ is smaller than ___.

4 Have students select one of their unit fractions to create a counting pattern with ten numbers, e.g. $0, \frac{1}{3}, \frac{2}{3}, 1, 1\frac{1}{3}$ etc.

5 Have students create three statements involving their unit fractions and a collection of 20 items, e.g. $\frac{1}{4}$ of 20 is 5. They may use counters or draw arrays to help.

Fractions and decimals
MA2-7NA represents, models and compares commonly used fractions and decimals

Year 3: Assessment Task Card

3.27

More About Fractions — TARGETED ASSESSMENT

If the student is experiencing difficulty:

Q2 Revise the fraction wall, comparing the relative size of each fractional part. Find the smallest and the biggest. Reinforce the idea that the more parts a whole is divided into, the smaller each part will be.

Q3 Have the student use the fraction wall to find fractions that 'line up' (are equivalent) or those that are bigger/smaller than the specified fraction.

Q4 Review the work on number lines from Lesson Plan 2. Have the student begin by using the fraction wall to decide how many parts make a whole, and use this knowledge to count to 1. Use the pattern created to continue the number line.

Q5 Revise the work on fractions of collections from Lesson Plan 3. Simplify the task by using a collection of 8. Have the student model with two lines of four counters, then create statements using $\frac{1}{2}$ and $\frac{1}{4}$.

Fractions and decimals
MA2-7NA represents, models and compares commonly used fractions and decimals

Year 3: Assessment Task Card

Decimals
Resources: A4 paper

1 Give each student a sheet of paper and have them divide their page into four sections.

2 In each section, have students write a number with one decimal place between 1 and 4, e.g. 1.7, 2.3, 3.8, 1.2.

3 In another section, have them write the equivalent mixed number (whole number and fraction) for each number.

4 In another section, have them draw a diagram to represent each number.

5 On the back of the sheet, have students order their numbers from smallest to largest.

6 Finally, have students construct a number line from 1 to 4 and place their numbers to scale on the number line.

Fractions and decimals
MA2-7NA represents, models and compares commonly used fractions and decimals

Year 3: Assessment Task Card

Decimals TARGETED ASSESSMENT

If the student is experiencing difficulty:

Q2 Revise what a decimal is. Review the work on tenths from Lesson Plan 1. Have the student list the whole numbers that fall between 1 and 4, then add in the tenths values.

Q3–4 Review the work on comparing fractions, decimals and diagrams from Lesson Plan 1. Have the student say what 'tenth' means (out of 10) and think of the fraction that corresponds to this.

Q5 Review the work on the place-value mat with tenths from Lesson Plan 2. Have the student begin by ordering the whole numbers, then drawing diagrams of the tenths parts to decide which number (with the same whole number) is bigger.

Q6 Revise the work on decimals on number lines from Lesson Plan 3. Have the student draw a line and write 1 and 4 at either end. Discuss how many whole numbers there are between 1 and 4 and mark partitions equally on the line. Look at the numbers in order from Q5 and decide on their position. Ensure that the student understands that the decimal 0.5 is equivalent to half, and uses this knowledge to help with placements on the number line.

Fractions and decimals
MA2-7NA represents, models and compares commonly used fractions and decimals

Year 3: Assessment Task Card

Quadrilaterals and Symmetry
Resources: A4 paper, BLM 79 '2D Shapes'

1 Give each student a sheet of paper. Have each student select three cards from BLM 79 '2D Shapes'. Have them identify each shape and write a description of it.

2 Present students with the following shapes and have them identify if they have a line of symmetry.

3 Present students with the following shapes and have them identify which ones have more than one line of symmetry.

4 Have students draw a square and all its lines of symmetry.

Two-dimensional space
MA2-15MG manipulates, identifies and sketches two-dimensional shapes, including special quadrilaterals, and describes their features

Year 3: Assessment Task Card

Quadrilaterals and Symmetry

TARGETED ASSESSMENT

If the student is experiencing difficulty:

Q1 Revise 2D shapes and their features. Have the student identify and name 2D shapes and discuss how many corners and sides they have. Have the student compare and contrast different 2D shapes and explain how they are similar and different. Have the student identify 2D shapes around the classroom.

Q2 Review what 'symmetrical' means and have the student fold or cut 2D shapes in half to recognise whether the two sides are identical. Provide opportunities for the student to use a mirror to explore symmetrical shapes. Have the student identify symmetrical objects or shapes in the room. Revise how the line in the middle of a symmetrical shape is called the line of symmetry.

Q3–4 Revise how some shapes have more than one line of symmetry. Have the student fold cut-outs of shapes in different ways to identify all the lines of symmetry.

Two-dimensional space
MA2-15MG manipulates, identifies and sketches two-dimensional shapes, including special quadrilaterals, and describes their features

Year 3: Assessment Task Card

Volume and Capacity

Resources: A4 paper, four different sized containers with different capacities (greater than 250 mL, but unmarked); a 250 mL container (unmarked; pop-top or cup would be ideal); measuring jug, interlocking centimetre cubes

1. Give each student a sheet of paper and four different containers (greater than 250 mL) and ask them to estimate the order of items from smallest to largest capacity. Have them record their order.
2. Using a 250 mL container, have students estimate how many of this container each of their four containers will hold. Have them record their estimates and explain their reasoning.
3. Have students measure how many of their 250 mL container will fit into their other containers and compare the actual result to their estimations.
4. Have students use measuring jugs to measure the amount of water held by each container and use this information to order the containers from smallest to largest capacity.
5. Show a model made from eight interlocking centimetre cubes. Ask students to describe it and to give the volume.
6. Ask students to use the interlocking centimetre cubes to make a model that has a a volume of 6 cm^3.

Volume and capacity

MA2-11MG measures, records, compares and estimates volumes and capacities using litres, millilitres and cubic centimetres

Year 3: Assessment Task Card

Volume and Capacity

TARGETED ASSESSMENT

If the student is experiencing difficulty:

Q1 Have the student look at the size and shape of each container to decide how much it will hold, comparing two items at a time.

Q2 Remind the student of the class activities from Lesson Plan 1, as well as 'How Many?' from Lesson Plan 2, Tuning In. Have the student look at the size of the 250 mL container compared to the others and predict how many times bigger the other containers are.

Q3 Emphasise the need for accurate measurement – encourage the student to fill the small container to the top and to pour without spilling into the larger container. Students may work together on this, so one can pour and the other can count.

Q4 Revise activities on using measuring jugs from Lesson Plan 3. Ensure that the student pours the water carefully and reads the scale accurately.

Q5–6 Show a single interlocking centimetre cube and establish the volume as one cubic centimetre. Build models one cube by one and identify the volume each time a cube is added.

Volume and capacity

MA2-11MG measures, records, compares and estimates volumes and capacities using litres, millilitres and cubic centimetres

Year 3: Assessment Task Card

3D Objects
Resources: A4 paper, set of 3D objects, playdough

1 Give each student a sheet of paper. Have them select three objects from a set of 3D objects and write the name of each object.

2 Have students write a description of each object by explaining its faces, edges and corners.

3 Have students draw the 2D shapes they would see if they were looking at the faces of a triangular prism.

4 Tell students that you are looking at a 3D object and you can see a square. Have them list the 3D objects you could be looking at.

5 Show students a pyramid from a set of 3D objects. Have them explain how they know it is a pyramid.

6 Show students a prism from a set of 3D objects. Have them explain how they know it is a prism.

Three-dimensional space
MA2-14MG makes, compares, sketches and names three-dimensional objects, including prisms, pyramids, cylinders, cones and spheres, and describes their features

Year 3: Assessment Task Card

3D Objects TARGETED ASSESSMENT

If the student is experiencing difficulty:

Q1–2 Revise 3D objects and their features. Have the student locate and name 3D objects and discuss their faces, edges and corners. Have them compare and contrast different 3D objects and explain how they are similar and different.

Q3–4 Review how the faces of 3D objects are 2D shapes. Have the student look at the faces of 3D objects and record the shapes they can see. Have them explore which 3D objects have the same shaped faces.

Q5 Review the features of pyramids. Present the student with a collection of pyramids and have them explain what they have in common. Encourage them to trace around the faces of pyramids and discuss what they notice.

Q6 Review the features of prisms. Present the student with a collection of prisms and have them explain what they have in common. Encourage them to trace around the faces of prisms and discuss what they notice.

Three-dimensional space
MA2-14MG makes, compares, sketches and names three-dimensional objects, including prisms, pyramids, cylinders, cones and spheres, and describes their features

Year 3: Assessment Task Card

Money
Resources: A4 paper, play money or BLM 22 'Banknotes' and BLM 87 'Coins'

1. Give each student a sheet of paper. Have them select an amount of money between $1 and $9 which contains both dollars and cents.
2. Have students model their amount using play money or BLM 22 'Banknotes' and BLM 87 'Coins'.
3. Have students model the same amount using a different set of notes and/or coins.
4. Have students work out the change they would receive from $10 if they spent their amount of money. Have them model it using play money or BLM 22 'Banknotes' and BLM 87 'Coins'.
5. Have students model the change a different way, this time using only 50 cent, 10 cent and 5 cent coins.
6. Show an item priced at $2.43. Ask, 'What coins could you use to pay for this without needing any change?'

Addition and subtraction (money)
MA2-5NA uses mental and written strategies for addition and subtraction involving two-, three-, four- and five-digit numbers

Year 3: Assessment Task Card

Money

TARGETED ASSESSMENT

If the student is experiencing difficulty:

Q1 Remind the student that money values are written to two decimal places. Remind them of the coins we use in Australia and have them make sure the amount ends in 5 or 0.

Q2 Revise the value of our coins and banknotes. Have the student practise counting techniques to assist their adding, such as making 10 or 100 or adding the largest numbers first.

Q3 Have the student look at the relationships between numbers and our coin and note values. Emphasise links, e.g. two $5 notes equals $10 and two 50 cent coins equals $1.

Q4–5 Revisit 'All About Change' from the Whole-Class Introduction and other activities from Lesson Plan 3. Have the student practise using either subtraction or the counting-up strategy to work out the difference between two values.

Q6 Choose a price that is close to a multiple of ten, such as 39 cents. Discuss the real price and the coins that could be used to pay for the item.

Addition and subtraction (money)
MA2-5NA uses mental and written strategies for addition and subtraction involving two-, three-, four- and five-digit numbers

Test A: Student Sheet

Name: ______________________ Date: ____________

Whole numbers

1 Draw a circle around the **odd** numbers.

1 038 999 102 3 294 417

2 Order the numbers from Question 1 from **smallest** to **largest**.

3 Write *four thousand, three hundred and sixty-seven* as a numeral.

4 **A** Make three different numbers from the digits below.

1 3 5 7

B Add 100 to each number and write the new numbers formed.

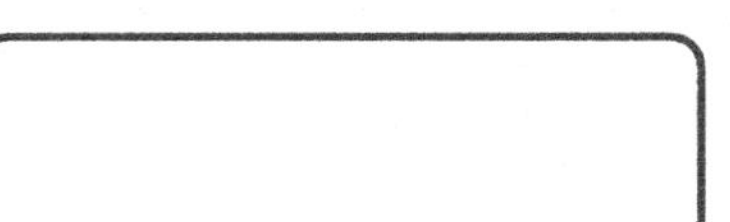

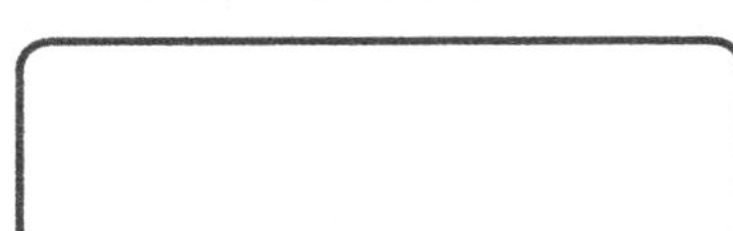

Addition and subtraction

5 Partition 2 560.

☐ +  + 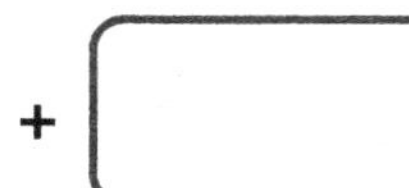 + ☐ = ☐

6 Write the value of the 3 in each of these numbers.

231 103 3 278 4 399

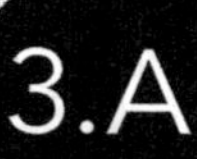

7 Add 3 000 to 4 369. ☐

8 Solve the following problems.

8 + 8 = ☐ 8 + 9 = ☐

Explain the mental strategy you used.

☐

9 Solve the following problem.

Jaxon had 39 marbles and Karla had 47 marbles. How many marbles did they have when they put their marbles together? ☐

10 If 49 - 7 = 42 then 39 - 7 = ☐

11

$$\begin{array}{r} 64 \\ -\ 27 \\ \hline \square \end{array}$$

12 Write the facts that can be made with the numerals: 4, 7 and 11.

☐ + ☐ = ☐ ☐ - ☐ = ☐

☐ + ☐ = ☐ ☐ - ☐ = ☐

Test A: Student Sheet

13 Solve the following problem.

Ki had $178 in the bank and $46 in his money box.
He bought a new bike for $89 and a helmet for $23.
How much money does he have left over?

14

Add 347 and 295.

Length

15 Circle the line that measures **exactly 5 cm and 3 mm.**

16 Circle the items that have a length that is **less than 1 metre**.

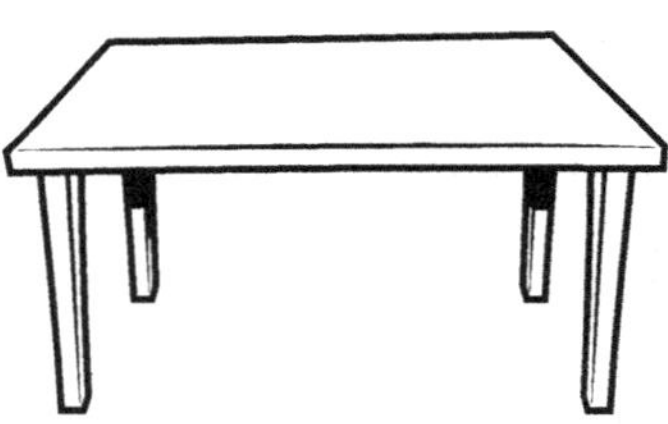

Mass

17 **A** Write the weight shown on the scale.

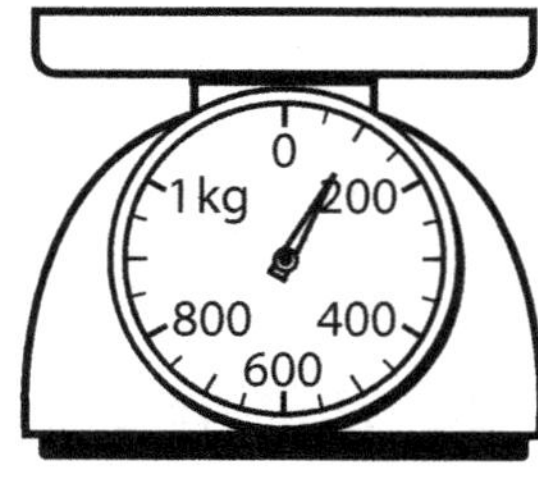

B Circle the item that would weigh **closest** to the amount shown.

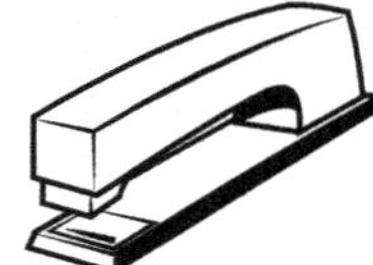

Test A: Student Sheet

3.A

Time

18 Write the time shown.

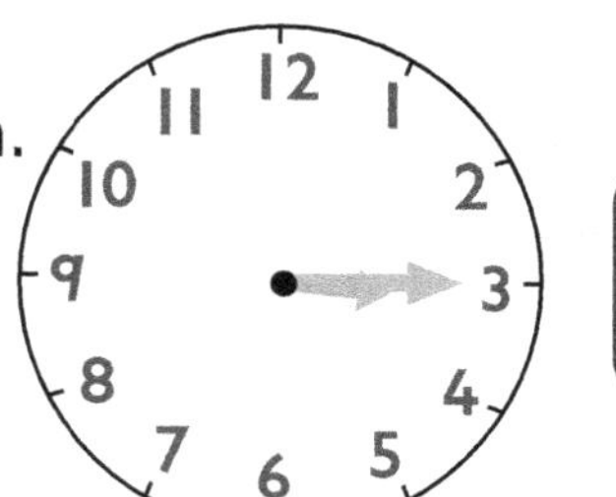

19 Draw hands on the clock to show 4:30.

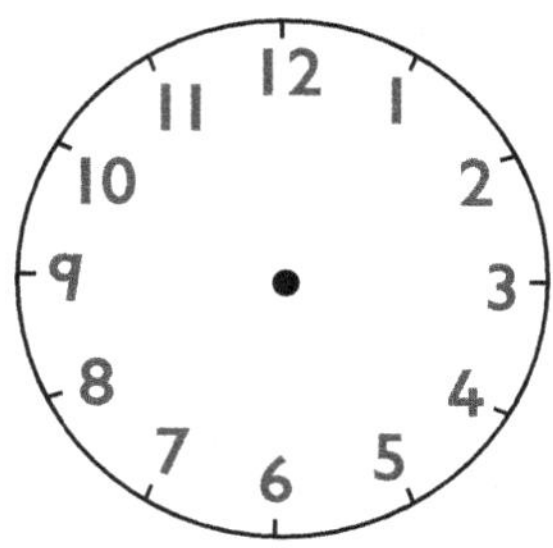

Position

20 Sara was at the entrance of the zoo. She wanted to look at the elephants and then at the monkeys.

A On the map, draw a path she could take.

B In the box, write directions explaining where she should go.

Test A: Student Sheet

3.A

Data

21 **A** Fill in the totals on the data collection table.

Ice-cream flavour	**Tally**	**Total**
Vanilla	III	
Chocolate	~~IIII~~ III	
Strawberry	~~IIII~~	
Banana	II	
Rainbow	~~IIII~~	

B How many students were interviewed altogether?

C Draw a column graph to represent your data.

Title: ______________________________

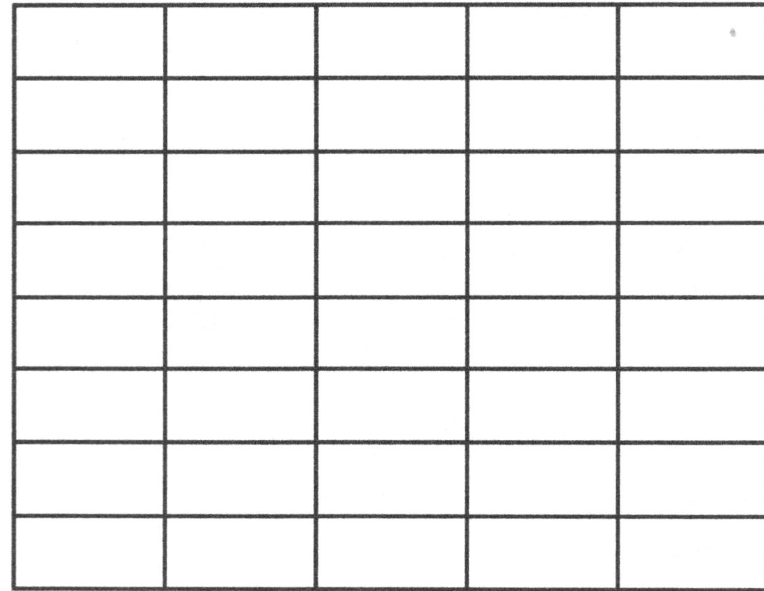

D Write two facts shown by the graph.

Test B: Student Sheet

3.B

Name: ______________________________ Date: ________________

Whole numbers

1 
How many MAB thousands would you need to make the following numbers?

6 934 852 1 099

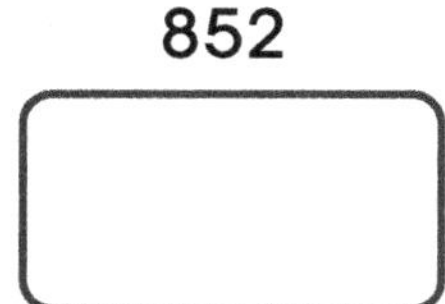

2 Calculate 10 less than these numbers.

6 934 852 1 099

3 Write 4 000 + 700 + 30 + 1 as a numeral.

4 **A** Use the following number expander to write the number 6 321.

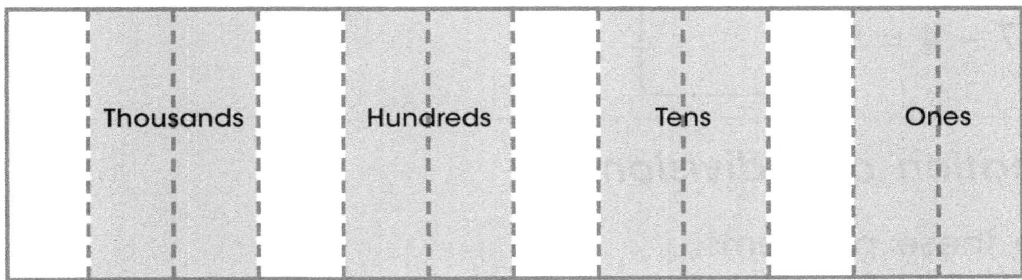

B Rename 6 321 with no hundreds.

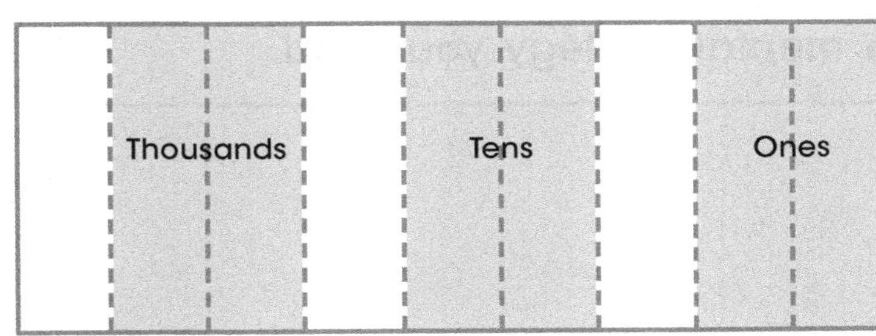

Test B: Student Sheet

3.B

Patterns and algebra

5 Create a number pattern to the rule:

Starting at 16, add 5.

Note: some spaces are filled in for you.

Addition and subtraction

6

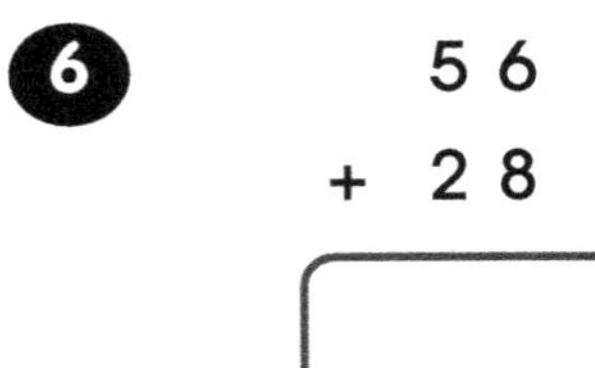

7 **A** Estimate the answer to the following problem.

79 – 32 ☐

B Circle the number that would be the correct answer.

59 34 47 63

8 Complete the number sentence.

17 – 4 = 9 + ☐

Multiplication and division

9 Solve these problems.

$6 \times 2 =$ ☐ $4 \times 2 =$ ☐ $7 \times 2 =$ ☐

Explain the mental strategy you used.

Test B: Student Sheet

10 Complete the table.

	6	3	8	2
x 5				

Explain the mental strategy you used.

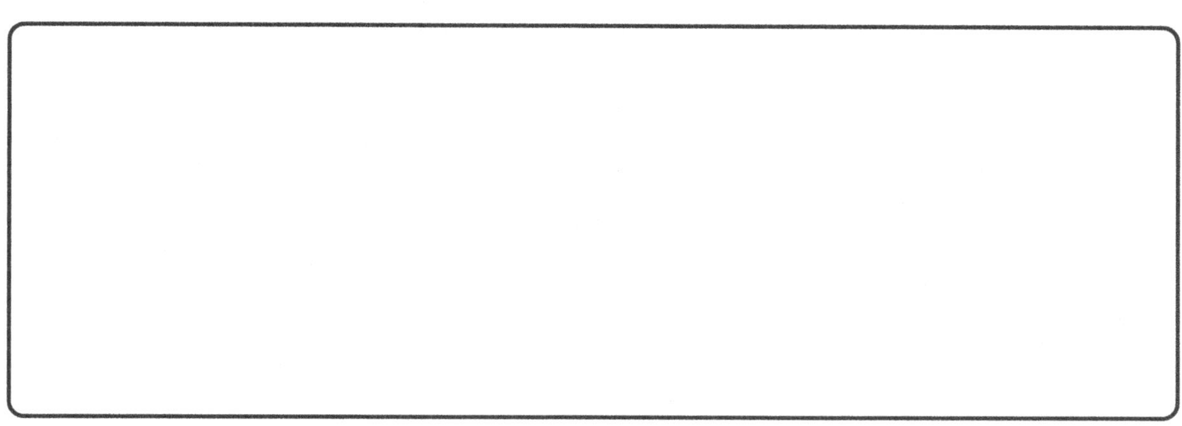

11 Write a multiplication problem to match this array.

12 Record the division problem and solve.

There were 24 strawberries in the box.
They were shared between the 8 children at the party.
How many strawberries did each child get?

13 Solve this problem.

27 ÷ 4 = ☐ Does it have a remainder? Yes/No

Test B: Student Sheet

14 Complete the problems.

If 3 × 5 = ☐ Then ☐ ÷ 3 = 5

Fractions and decimals

15 Show three ways to divide squares into two halves.

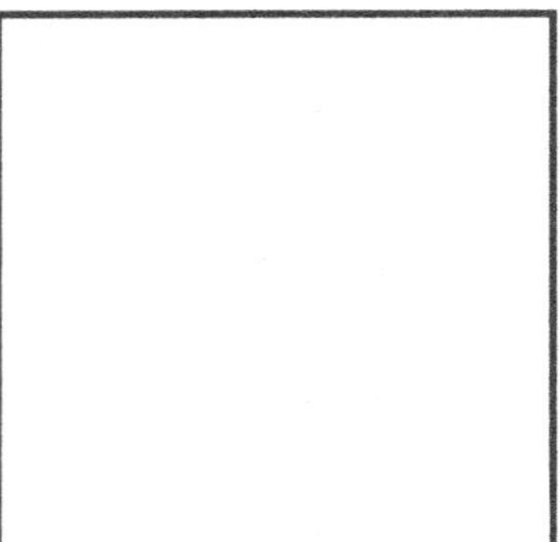

16 Write the fraction that is shown.

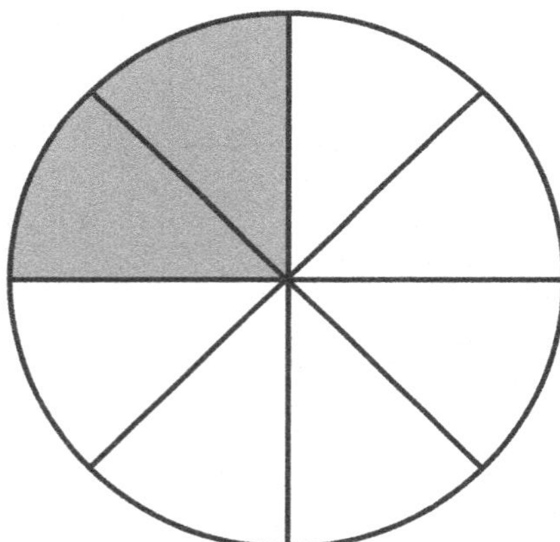

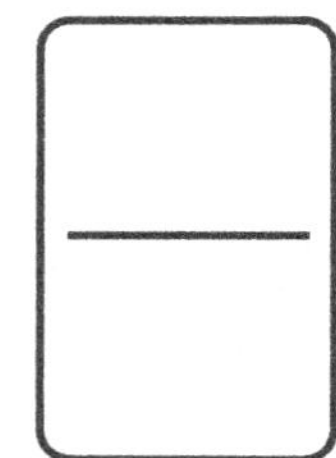 or ☐

17 Complete the fraction counting pattern.

$\frac{1}{4}$, $\frac{1}{2}$, $\frac{3}{4}$, 1, $1\frac{1}{4}$, ☐, , , ☐

Test B: Student Sheet

3.B

18 What is one third of 18 apples?

19 **A** Shade the shape to show $\frac{5}{10}$.

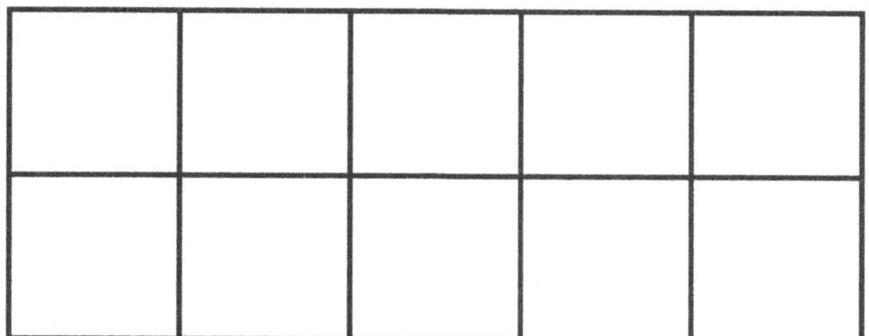

B Write this fraction as a decimal.

20 Look at the numbers below.

A Circle the **largest** number.

B Put a line under the **smallest** number.

0.7 1.2 0.1 3.1 5.6 0.8

Addition and subtraction

21 Write the **total value** of the following coins.

Test B: Student Sheet

3.B

Time

22 Draw hands on the clock to show 6:50.

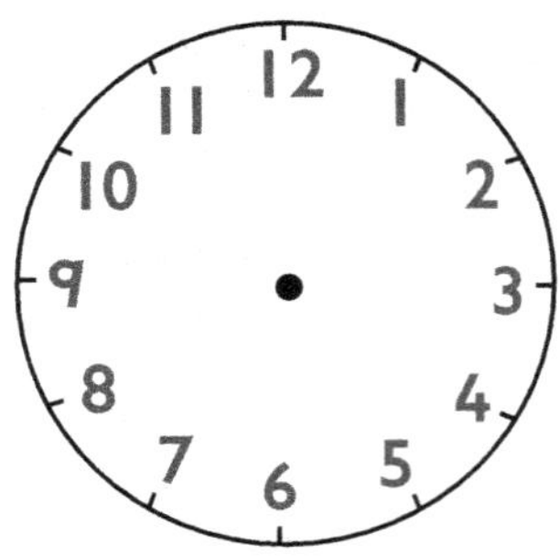

Area

23 Draw a shape with an area of exactly 10 cm^2.

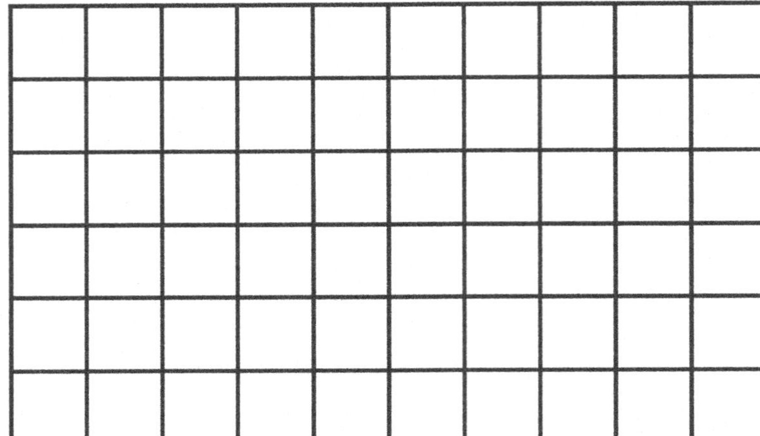

Volume and capacity

24 **A** Write the amount of water in the measuring jug.

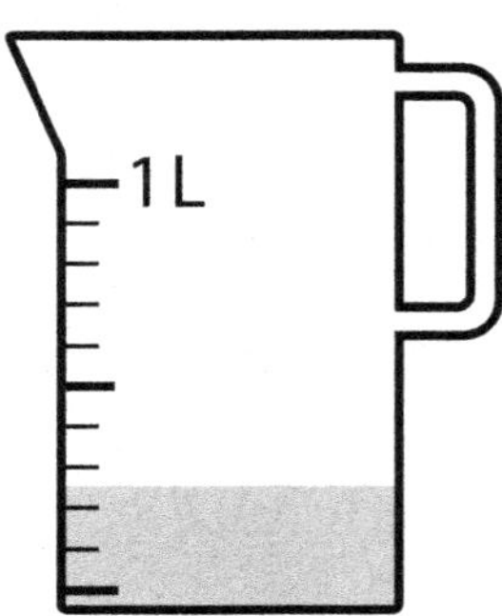

B What is the volume of this model? It is made from centimetre cubes.

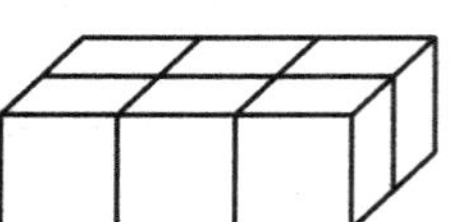

Test B: Student Sheet

3.B

Chance

25 Order the events from **least likely** (1) to **most likely** (3).

A I will eat lunch today

B A circus will come to school today.

C It will rain today.

1 **3**

Angles

26 Look at angles in the picture.

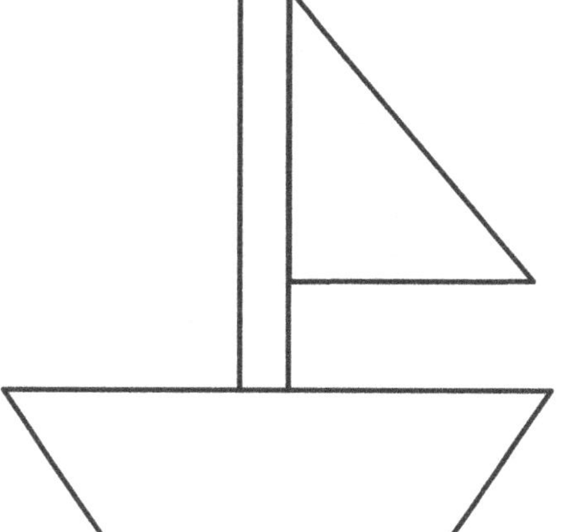

A A Put a star in the angles that are **smaller than** a right angle.

B Put a tick in the angles that are **bigger than** a right angle.

C Put a dot in the angles that are **exactly** a right angle.

Two-dimensional space

27 Circle the symmetrical shape.

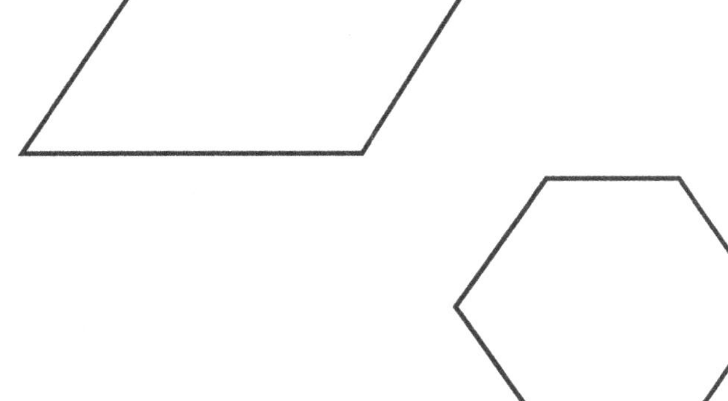

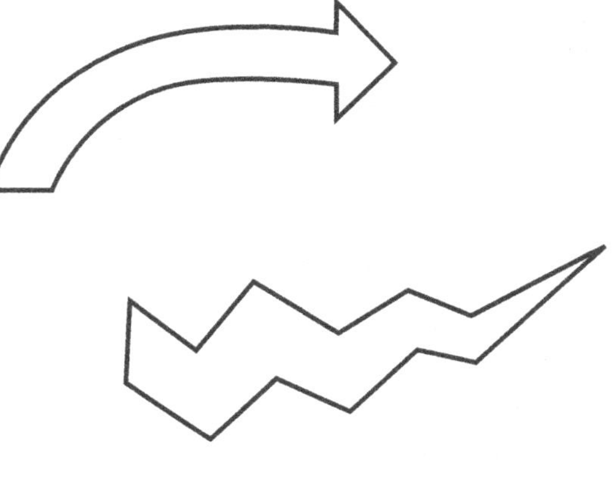

Three-dimensional space

A Name and describe this 3D object.

B Name and describe this 3D object.

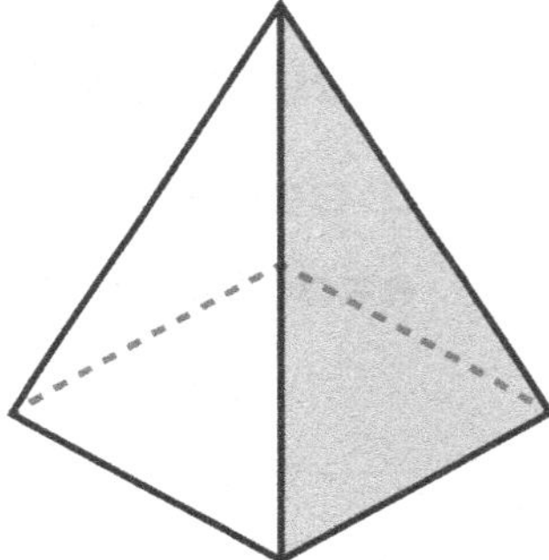

Class Unit Overview

Class: ________ Year: ________

32																								
31																								
30																								
29																								
28																								
27																								
26																								
25																								
24																								
23																								
22																								
21																								
20																								
19																								
18																								
17																								
16																								
15																								
14																								
13																								
12																								
11																								
10																								
9																								
8																								
7																								
6																								
5																								
4																								
3																								
2																								
1																								
Initials																								

Class SB Assessment Overview Class: ________ Year: ______

32/p. 131																								
31/p. 127																								
30/p. 123																								
29/p. 119																								
28/p. 115																								
27/p. 111																								
26/p. 107																								
25/p. 103																								
24/p. 99																								
23/p. 95																								
22/p. 91																								
21/p. 87																								
20/p. 83																								
19/p. 79																								
18/p. 75																								
17/p. 71																								
16/p. 67																								
15/p. 63																								
14/p. 59																								
13/p. 55																								
12/p. 51																								
11/p. 47																								
10/p. 43																								
9/p. 39																								
8/p. 35																								
7/p. 31																								
6/p. 27																								
5/p. 23																								
4/p. 19																								
3/p. 15																								
2/p. 11																								
1/p.7																								
Initials																								

Class Assessment Task Card Overview

Class: ________ Year: ________

Initials	3.1	3.2	3.3	3.4	3.5	3.6	3.7	3.8	3.9	3.10	3.11	3.12	3.13	3.14	3.15	3.16	3.17	3.18	3.19	3.20	3.21	3.22	3.23	3.24	3.25	3.26	3.27	3.28	3.29	3.30	3.31	3.32

WEEKLY MATHS PLANNER

Unit: ______ Week: ______ Term: ______ Date: ______ Year Level: ______

Resources: ____

		Monday	Tuesday	Wednesday	Thursday	Friday
	Tuning In					
	Whole-Class Introduction					
Small Group Focus	Independent Tasks (individual, pair, small group)					
Small Group Focus	Teaching Group					
	Reflection					

TERM MATHS PLANNER

Year Level: ________________ Term: ________________ Year: ________________

Week	Unit/page no.	Strand/Sub-strand/Content description/Code

YEARLY MATHS PLANNER

Year: ________________ Year Level: ________________

Term 1	Unit
Week 1	
Week 2	
Week 3	
Week 4	
Week 5	
Week 6	
Week 7	
Week 8	
Week 9	
Week 10	

Term 2	Unit
Week 1	
Week 2	
Week 3	
Week 4	
Week 5	
Week 6	
Week 7	
Week 8	
Week 9	
Week 10	

Term 3	Unit
Week 1	
Week 2	
Week 3	
Week 4	
Week 5	
Week 6	
Week 7	
Week 8	
Week 9	
Week 10	

Term 4	Unit
Week 1	
Week 2	
Week 3	
Week 4	
Week 5	
Week 6	
Week 7	
Week 8	
Week 9	
Week 10	